MINDS AND COMPUTERS

for G
who Helped

and

for Sue
without whom . . .

MINDS
AND
COMPUTERS

AN INTRODUCTION TO THE
PHILOSOPHY OF ARTIFICIAL
INTELLIGENCE

Matt Carter

EDINBURGH UNIVERSITY PRESS

Edinburgh University Press Ltd
22 George Square, Edinburgh

Typeset in Times
by Servis Filmsetting Ltd, Manchester, and
Printed and bound in Great Britain by
CPI Antony Rowe, Chippenham and Eastbourne
Transferred to Digital Print 2009

A CIP record for this book is available
from the British Library

ISBN 978 0 7486 2098 2 (hardback)
ISBN 978 0 7486 2099 9 (paperback)

CONTENTS

ACKNOWLEDGEMENTS

I'd like to express my gratitude to all who participated, directly and indirectly, in the production of this book.

Thank you to the teaching staff who were based in the Cognitive Science programme at the University of Queensland in the final years of the twentieth century, for inspiring in me a commitment to the importance of cross-disciplinary analysis. My gratitude further extends to all of my teachers – both within and without Philosophy.

Thank you also to the years of undergraduates who have suffered my instruction. In particular, I am grateful to my 'Minds and Machines' class of 2002 for inspiring this textbook in the first instance, and to my classes of 2006 for reading and commenting on material contained herein.

Thank you to the Philosophy Department at Melbourne University, where I was based while writing this book, and to its superb office staff.

Thank you to all at EUP for publishing this volume and for being such a pleasure to deal with. Particular thanks to Jackie Jones for her initial enthusiasm for the project.

Thank you to all my friends for their support and understanding, particularly to FB, Wayne and Eloise for tolerating innumerably many dinner-time drop-ins, and to Lester and Christie for assistance above and beyond the call of friendship.

Thank you very much to Graham Priest, without whom this book would not have been written.

Thank you to Mia and Linus for being adorable, and a million thank yous to Sue, for being wonderful.

CHAPTER 1

INTRODUCTION

This is a book about minds. It is also about computers. Centrally, we will be interested in examining the relation between minds and computers.

The idea that we might one day be able to construct some artefact which has a mind in the same sense that we have minds is not a new one. It has featured in entertaining and frightening fictions since Mary Shelley first conceived of Frankenstein's monster.

In the classic science fiction of the early to mid-twentieth century, this idea was generally cashed out in terms of 'mechanical men' or *robots* – from the Czech word *robata*, which translates roughly as the feudal term *corvée*, a term which refers to the unpaid labour provided to one's liege lord.

In more modern fiction, the idea of a mechanical mind has given way to the now commonplace notion of a computational *artificial intelligence*. The possibility of actually developing artificial intelligence, however, is not just a question of sufficiently advanced technology. It is fundamentally a *philosophical* question.

It is this question that we will be centrally concerned with throughout this volume. In order that we might be in an informed position to consider the possibility of artificial intelligence, we will need to answer a number of related questions.

Firstly, we will be asking just what the human mind is. The twentieth century saw a succession of philosophical theories of mind, culminating in the currently dominant theory which accommodates the possibility of artificial intelligence. Our first goal, which we will spend Chapters 1 to 10 pursuing, is to clearly articulate this theory.

Philosophically responsible engagement with this theory requires a sound understanding of precisely what a computer is. Consequently, we're going to spend three chapters developing a rigorous technical account of computation. Although this material is technical, the

1

introduction is slow and gentle and will be readily accessible to a reader with no background in mathematics or computer science.

Along the way to our target theory, we're going to survey the space of available philosophical theories of mind, weighing the merits and flaws of each. This will provide a comprehensive introduction to the philosophy of mind.

We are also going to take a couple of empirical diversions along the way. We're going to tell the story of the rise of empirical psychology and we will spend a chapter developing a rudimentary understanding of functional neuroanatomy.

Once we are armed with a sound philosophical understanding of our target theory, the remainder of the book will be given to evaluating it. We will see that a wide range of material from the empirical disciplines bears importantly on the tenability of the theory. As such, this book is overarchingly an exercise in cross-disciplinary analysis.

We're going to focus on two mental capacities that are distinctively human – our capacity for reasoning and our facility for language. Our aim first of all in Chapters 11 to 20 will be to compare what we know of the human rational and linguistic capacities with methods for implementing these computationally.

We will see how computers can be programmed for strategic game play, for reasoning about novel situations based on known information and for certain functions implicated in language production and comprehension. This will expose us to some introductory material in linguistics and a tiny bit of formal logic, and we will touch on some material from cognitive psychology.

In the final chapters of the book we will examine some more advanced philosophical material concerning the notions of meaning and representation. Lastly, we will introduce artificial neural networks and see how they can be employed in the pursuit of artificial intelligence – again with particular respect to rational and linguistic functions.

All told, we will be examining material from philosophy, psychology, linguistics, neuroscience and computer science – the disciplines which constitute cognitive science. It is to be expected that most readers will find some of this material more approachable and some less so; however, I have aimed for maximal accessibility to the introductory reader throughout.

Each chapter from Chapter 11 onwards engages with an issue which by all rights deserves a dedicated volume. As such, the coverage is less than comprehensive and I have frequently simplified explanations in the name of accessibility. There are suggestions for further reading at the end of the book for readers who want to further their understanding of the issues we cover.

Comprehensive coverage of the relevant issues, however, is not our primary concern here. Our main aim is to develop and evaluate the philosophical theory of mind which allows for the possibility of artificial intelligence. By the end of the book, the reader should find themselves in a sound position from which to make informed decisions concerning the possibility of developing artificial intelligence. This book should also provide a solid foundation for philosophically responsible engagement with cognitive science broadly.

We're now going to begin our tour of the space of available philosophical theories of mind with a theory which most people implicitly, and pretheoretically, subscribe to: dualism.

CHAPTER 2

DUALISM

We are going to begin our examination of the available theories of mind with Cartesian dualism. There are at least two good reasons for doing so. One is that the presentation of theories of mind in the following chapters will be broadly chronological and – at least as far as modern philosophy is concerned – to begin with Cartesian dualism is to begin at the beginning.

Another reason is that, by and large, people's pretheoretic intuitions concerning the mind and the body are dualist in general and Cartesian dualist in particular. Once you have read this chapter, ask your friends and family about their intuitions and I strongly suspect you will find as much.

Unfortunately, although a good starting place, Cartesian dualism is beset with philosophical difficulties. Let us then begin the task of making clear precisely what the Cartesian dualist is committed to and what the problems with the theory are.

2.1 SUBSTANCE DUALISM

Substance dualism is a metaphysical view. It is the view that the universe consists of two different kinds of stuff – two metaphysically distinct substances. As such, substance dualism is a commitment to a particular *ontology* – an ontology which sees the universe as comprised of both material and immaterial substances. As well as all the material stuff which makes up the physical world, the dualist holds that there is also non-physical, immaterial stuff to be taken account of.

We need to be careful in drawing the distinction between the material and the immaterial. For instance, electromagnetic radiation, while in a certain sense insubstantial, is still material. It is part of the physical world – something we would expect physics to give us an account of. The distinction between material and immaterial is not just a straightforward distinction between things we can bump into in the

4

dark (chairs and tables) and things we cannot (heat, light and sound). Rather, it is a distinction between the things which are within the purview of physics (chairs, tables, heat, light, sound) and the things which the dualist contends exist beyond the scope of physics.

This is difficult to grasp in the abstract so let's examine a particular kind of substance dualism and see why we might be tempted to claim that there are objects in the universe composed of non-physical, immaterial stuff.

2.2 CARTESIAN DUALISM

Cartesian dualism is a view about the mind, the body and the relation between them. It is a particular kind of dualism which takes its name from its original proponent, René Descartes. It is essentially the view that while the body is a material object, the mind is not. According to the Cartesian dualist, the mind is composed entirely of immaterial stuff. As such, Cartesian dualism is clearly a kind of substance dualism as the Cartesian dualist is committed to an ontology which admits of both material and immaterial substances.

What is distinctive about Cartesian dualism among other kinds of mind–body dualism is that the Cartesian dualist holds that the mind and the body enter into causal relations with each other: the mind causes things to happen in the body and the body causes things to happen in the mind. In other words, the immaterial mind and the material body *interact*. Cartesian dualism is also known as interactionist dualism for this reason.

The Cartesian dualist is committed to the following four propositions:

[D1]   The body is composed entirely of material substance.
[D2]   The mind is composed entirely of immaterial substance.
[D3]   The body has a causal effect on the mind.
[D4]   The mind has a causal effect on the body.

As we shall shortly see, it is very difficult to maintain all four of these propositions. But before we start examining objections to Cartesian dualism, let's first consider the arguments in its favour.

2.3 POSITIVE ARGUMENTS FOR CARTESIAN DUALISM

There are a number of reasons why one might endorse Cartesian dualism. I will consider the three strongest arguments in its favour: the argument from religion, the argument from introspective appearance and the argument from essential properties.

2.3.1 THE ARGUMENT FROM RELIGION

This is perhaps the most commonly held argument in favour of Cartesian dualism.

Many religions – Christianity amongst them – posit an afterlife and promise a reward in the afterlife for living according to a certain normative code. Conversely, they threaten punishment in the afterlife for failing to live according to these dictates. But ask yourself: who is it that is to be rewarded or punished?

Such religions speak of the eternal, immutable, immaterial soul which is contended to be, in an important sense, constitutive of the individual. It is this eternal soul which enjoys the rewards or suffers the punishments meted out in the afterlife.

In order for the concepts of reward and punishment to be applicable – and in order for the relevant beliefs to motivate individuals to act in the appropriate way – it must be the case that the thing that is rewarded or punished is the same thing that is responsible for moral agency.

In other words, the thing which is rewarded or punished simply *must* be the same thing that goes about in the world making decisions and acting in certain ways. The soul simply must be equivalent to the mind. After all, what sense lies in rewarding or punishing one entity for the deeds or misdeeds of a distinct entity? And why should I be at all concerned with acting according to a particular code if it is not, in a very important sense, *me* who will enjoy the promised reward or suffer the threatened punishment?

Cartesian dualists understand 'mind' and 'soul' to be synonymous terms. While in life, the mind/soul stands in relation to a particular body. In the afterlife, the immaterial mind/soul leaves the body to take up independent existence and to enjoy or suffer the rewards of the actions it engaged in during its materially embodied life.

So, to the extent to which one is antecedently committed to such a religious doctrine, one must also be committed to substance dualism as such doctrines require an immaterial soul. Furthermore, one must also be committed to Cartesian dualism as it must be the case that the mind/soul *causes*, and is thereby responsible for, the actions of the body.

As far as arguments go, unfortunately, this is not a very good one. It gives no independent reason whatsoever for endorsing Cartesian dualism. What it shows is that a commitment to Cartesian dualism is a straightforward corollary of certain religious beliefs. This simply means that these religious beliefs stand or fall together with Cartesian dualism. If one doesn't antecedently have such religious beliefs, the argument from religion is entirely lacking in persuasive force.

2.3.2 THE ARGUMENT FROM INTROSPECTIVE APPEARANCE

Another argument in favour of dualism proceeds from our privileged introspective awareness of our own minds.

It is a distinctive feature of our minds that they have a reflective capacity: we can think about our own thoughts, our own mental states. Furthermore, we have *unique* and *privileged* access to the contents of our own mental states. This access in unique in that I and I alone am privy to *my* mental life. It is privileged in that, unlike my access to everything else in the universe, my access to my own mental life is *direct* and not mediated by my senses.

Given this capacity of minds for, and amenability of minds to, direct introspection, we might be tempted to draw a distinction between minds and physical objects, as follows.

When I introspect – when I reflect on my mental life and consider the contents of my mental states – it doesn't *seem* to me that events in my mental life are physical events. My thinking of ice cream seems to me to be just that – thinking of ice cream. It doesn't seem at all to be an electrochemical discharge in my brain. It doesn't seem at all to be *any* kind of physical event. So my thinking of ice cream must be a *non-physical* event – mentality must be a non-physical phenomenon, in which case we are committed to substance dualism at the least.

The *argument from introspective appearance*, as given in the previous paragraph, suffers a rather tenuous connection between its premises – which are concerned with the way things *seem* to us – and its conclusion, which maintains that the way things *are* is in accord with the way things *seem*. The obvious reply, then, to the argument from introspective appearance is to point out that there is no necessary connection between the way things appear to us and the way things actually are. In fact, we are very often deceived by the way things seem: a warm breeze doesn't *seem* at all like the kinetic energy of millions of molecules, nor does electricity *seem* at all like a flow of electrons.

The way things *seem*, however, in no way establishes a distinction between a warm breeze and the kinetic energy of millions of molecules on the one hand, or between electricity and a flow of electrons on the other. Similarly, the way our minds *seem* to us cannot be relied upon to establish the non-physicality of mentality.

The fallibility of appearances certainly does not entail that mentality *cannot* be non-physical. It merely shows that the argument from introspective appearance fails as an argument in favour of the non-physicality of mentality as it fails to establish the required necessary connection between the truth of its premises and the truth of its conclusion.

2.3.3 THE ARGUMENT FROM ESSENTIAL PROPERTIES

The argument from introspective appearance appealed to the fact that minds have certain essential properties which ordinary physical objects lack, and vice versa. For one thing, minds have this essential capacity for direct introspection and reflective awareness that ordinary physical objects do not have. But minds and physical objects also differ in other essential ways.

Ordinary physical objects are essentially publicly accessible (anyone can observe a chair) whereas minds are not (only I can directly observe my mind). Ordinary physical objects are also essentially extended in space – they have mass, shape, location and other spatial properties. Minds, on the other hand, are essentially thinking things: they don't, merely by virtue of being minds, have spatial properties in the way that a chair, merely by virtue of being a chair, has spatial properties. The only properties minds have merely by virtue of being minds are those pertaining to capacities to think. It is not an essential property of minds that they are extended in space in the way that it *is* an essential property of ordinary physical objects, such as chairs, that they be spatially extended.

Given this radical divergence in essential properties between minds and ordinary physical objects, there must, then, be a *distinction in kind* – minds must be a different *kind* of thing to physical objects. They must, therefore, be non-physical entities.

While the *argument from essential properties* we have just rehearsed seems initially compelling, a little thought serves to dispel its force. There is much we could say about the metaphysics of essential properties which the argument trades on. For present purposes, however, it suffices to recognise that a distinction in kind is *not* tantamount to a metaphysically substantive distinction.

In other words, we can adopt the same strategy we employed in defusing the argument from introspective awareness – we concede the truth of the premises but point out that this does not establish a necessary connection to the truth of the conclusion. It is a given that minds are quite unlike anything else we know of in the universe. This does not, however, entail that minds are *made of different stuff* to everything else in the universe – i.e. that minds are composed of *non-physical* substance.

As before, the failure of the argument from essential properties does not entail that minds *cannot* be non-physical – it merely shows that a radical distinction in properties (essential or otherwise) between minds and canonical physical objects is not sufficient to establish that minds are non-physical.

2.4 ARGUMENTS AGAINST CARTESIAN DUALISM

We have now seen three arguments in favour of dualism. The argument from introspective appearance and the argument from essential properties both seek to establish a broad mind-body dualism. Coupled with certain common-sense intuitions concerning the efficacy of the mind in bringing about changes in the body and vice versa, they become arguments supporting Cartesian dualism. The argument from religion seeks to establish Cartesian dualism in particular, as the interaction between mind and body is essential for the argument.

We should already be concerned for the theory, given that we have actively sought the strongest arguments in its favour and have discovered that none of them succeed in establishing their conclusions. Even more troubling for the Cartesian dualist though are the following negative arguments.

2.4.1 THE PROBLEM OF OTHER MINDS

The first objection to Cartesian dualism we will consider identifies a problematic consequence of the view.

We have become quite adept at investigating the physical universe and have all manner of methods and equipment at our disposal for doing so. We are at a loss, however, when it comes to investigating the non-physical.

If minds are immaterial, then they are clearly not investigable by known empirical methods. Not only does this put minds beyond the scope of science, it also means that there is no way to know whether or not *other people* have minds. As far as the Cartesian dualist is able to discern, she may well have the only mind in the universe – all other human bodies may well just be mindless automata.

This is an *epistemological* concern – a concern about what we can know – which comes with a *methodological* concern for the possibility of a science of mind. The objection is not insuperable, however. The Cartesian dualist can help herself to a reply, of sorts, to each of these concerns.

With respect to scientific methodology, she might point out that it is not unknown to science to postulate, and investigate, unobservable entities by examining their observable consequences. So, while she will have to maintain that minds are simply not amenable to *direct* empirical investigation, she can hold out hope that there will be observable consequences of mentality that *can* be investigated, thereby giving science *indirect* access to minds.

With respect to our everyday knowledge of the minds of others, the Cartesian dualist can reason by analogy to her own mental life and its role in mediating experience and behaviour. Presumably – she might say – you do, in fact, think that other people have minds (it is fairly difficult to get around in the world without proceeding on that assumption). Why then do you think this? Presumably because you've observed that the best explanation for the way other people behave involves attributing mental states to them.

In other words, you know that if *you* have certain experiences, this will lead to certain *beliefs* and *desires* (mental states) which in certain situations will lead you to behave in particular ways. You've further observed other people in just such situations acting in just such ways and consequently assume that they share certain of your beliefs and desires (like the belief that it is lunchtime and the desire for food) and recognise that these mental states play an important explanatory role in understanding their behaviour.

This reply to the problem of other minds appeals to an *inference to best explanation*: the best way to explain the way other human bodies move around in the world is to attribute to them the kind of mental states I *know* that I have. It should be apparent, however, that while this reply suffices to demonstrate the utility in *assuming* that other people have minds (the assumption confers useful predictive capacities), it certainly does not *establish* that they do. The problem of other minds remains for the dualist.

2.4.2 OCKHAM'S RAZOR

William of Ockham was a medieval philosopher and notable logician of the early fourteenth century. You may well have heard a common corruption of Ockham's razor that is something along the lines of 'the simplest explanation is often the best'. Properly construed, Ockham's razor is intended to serve as a methodological constraint on theory construction.

The most accurate gloss of Ockham's razor in the realm of meta-physics is 'don't expand your ontology beyond necessity'. Another way of putting it is to say that one shouldn't postulate any more entities than are absolutely necessary to explain the phenomena about which we are theorising.

This can be deployed as a methodological objection to Cartesian dualism – the contention being that the dualist *does*, in fact, expand her ontology beyond explanatory necessity, that postulating non-physical entities is not required in order to explain mentality.

This a moderately weak objection so I shall give it short thrift. At best – *if* you think that the principle should constrain theory construction – it entails that when presented with two explanatorily adequate theories of mind, one of which postulates non-physical entities and one of which accounts for mentality in purely physical terms, one should prefer the latter. This will be something to bear in mind once we have surveyed the space of available theories of mind.

2.4.3 THE PROBLEM OF INTERACTION

A considerably more potent objection – one which is generally considered to be the rock on which Cartesian dualism founders – centres on the problem of interaction.

The physical universe is held to be *causally closed*, which means that every physical effect has a physical cause. A physical effect brought about by a non-physical cause would contravene the first law of thermodynamics. While science has certainly got it wrong about many things in the past, our theory of thermodynamics is a foundational theory which most of modern science rests on.

The problem here for the Cartesian dualist, if it is not already apparent, is their contention that the non-physical mind is causally efficacious in the physical world, that the non-physical mind *causes* change in the physical body.

What might the Cartesian dualist say to the problem of interaction? The only possible response seems to be to deny that the physical universe is, in fact, causally closed. This, however, seems rather implausible. Were it the case that our physical actions were *caused* by non-physical minds, then energy would be added to the physical universe every time a mental action resulted in a physical action and this addition of energy would, one might think, be measurable.

There is a theistic response available here, which is to claim that in *every* case of scientific observation, an omniscient, omnipotent divinity intervenes and adjusts the observer accordingly, such that we *think* that energy is always conserved and that the amount of energy in the physical universe is constant, but *in fact* it is constantly increasing. Taking such a line, however, brings with it a raft of troubling epistemological concerns.

There seems to be no secular way to rescue Cartesian dualism from this objection. We can, however, advance modified forms of dualism which retreat from the commitment to interaction.

2.5 OTHER DUALISMS

Recall from section 2.2 the four propositions [D1]–[D4] which characterise Cartesian dualism. One way to recover the core ontological intuitions of Cartesian dualism from the damning criticism of the problem of interaction is to give up the commitment to propositions [D3] and [D4], leaving us in want of an account of the relation between the physical body and the non-physical mind. This strategy leads to the theistic dualist theories known as parallelism and occasionalism.

Another possible strategy is to give up only [D4] and maintain a commitment to propositions [D1]–[D3]. Again, this requires a particular account of the relation between mind and body which is supplied by a theory known as epiphenomenalism. Let's deal with these other dualist theories seriatim.

2.5.1 PARALLELISM

Parallelism is a theistic dualist theory. The parallelist maintains the ontological independence of mind and body but denies that they interact causally. Once we deny causal interaction between the material body and the immaterial mind, we need to explain the apparent interaction in some other way. The parallelist appeals to an omniscient, omnipotent being to account for the connection.

According to the parallelist, when this omniscient, omnipotent being (henceforth simply 'God') created the physical universe of bodies and the non-physical universe of minds, She set things up in such a way that although the sequence of events in the physical universe is entirely causally independent of the sequence of events in the non-physical universe, the two sequences are in perfect harmony. This preordained harmony accounts for the correlation between my mental life and my physical life. Whenever I put my hand on a hot stove, I have the mental experience of the hurtfulness of pain because God set things up initially such that it would be so. Whenever I have the mental experience of desiring to move my arm, the physical event of my arm rising obtains because God set things up initially such that it would be so. And so forth.

Although parallelism circumvents the problem of interaction by denying any *causal* connection between material bodies and immaterial minds, it is still beset with the other objections which plague

Cartesian dualism. We are still stuck with the epistemological difficulty of a lack of access to other minds and we are still in contravention of Ockham's razor. Furthermore, the theory seems no more nor less plausible than our proposed Cartesian theistic response to the problem of interaction.

2.5.2 OCCASIONALISM

As with parallelism, occasionalism is a theistic dualist theory which denies interaction between the material body and the immaterial mind and appeals to a God to explain the connection between the two.

The only difference, in fact, between parallelism and occasionalism is that where the former holds that God set up the series of physical events and the series of non-physical events in preordained harmony, the occasionalist holds that God steps in where and as required in order to maintain the harmony of the two series.

According to the occasionalist then, whenever I put my hand on a hot stove, I have the mental experience of the hurtfulness of pain because God intervenes to ensure that it will be so. Whenever I have the mental experience of desiring to move my arm, the physical event of my arm rising obtains because God intervenes to ensure that it will be so. And so forth.

The particular doctrinal considerations which motivate the departure from parallelism are not of import here. For our purposes it suffices to observe that occasionalism enjoys the same benefits and suffers the same criticisms as parallelism.

2.5.3 EPIPHENOMENALISM

A notably more convincing dualist view is that of the epiphenomenalist. In fact, in contrast to the other dualist views we have covered, you will find numerous epiphenomenalists working in contemporary cognitive science.

The epiphenomenalist maintains propositions [D1]–[D3], rejecting only [D4]. She maintains the ontological distinction between the mental and the physical and *also* maintains the causal relation between the material body and the immaterial mind, but *only in one direction*. She asserts that physical states give rise to mental states, but *denies* the problematic converse.

The epiphenomenalist does not run afoul of the problem of interaction as she *does* maintain the causal closure of the physical universe. According to epiphenomenalism, every physical event is *wholly* and *solely* accounted for by antecedent physical events. In other words, every physical effect has a physical cause. As well as having physical effects, though, some physical states are held to also give rise to mental states.

The epiphenomenalist picture, then, is one of a chain of physical causation containing some physical states which also give rise to mental states. These mental states are held to be causally inefficacious – they don't *do* anything. Mental states, on this view, are mere *epiphenomena* of physical states. They are accounted for by physical states but they themselves cause neither physical states nor further mental states.

It is likely to be less than clear to you why one might want to maintain a theory which sees mental states as ontologically distinct from their associated physical states, yet causally inefficacious in both the physical and the mental realms. This is likely to become a lot clearer later in this volume when we discuss the privileged first-person experience of *having* or *being in* a mental state. For now, it is only important that you understand the mechanics of the theory and the way in which it differs from the other kinds of ontological dualism we have examined.

2.6 ANOMALOUS MONISM

A final theory we should at least mention before concluding this chapter is anomalous monism, otherwise known as double aspect theory or, simply, property dualism.

Anomalous monism is not, strictly speaking, a dualist theory in the sense of each of the other theories in this chapter, since the anomalous monist is *not* a substance dualist. The dualism they advance – such as it is – is a dualism of *properties*, not substances.

According to the anomalous monist, there is no non-physical *substance*. There are however, they contend, *irreducibly non-physical properties* of physical substance. In other words, certain physical states have a *double aspect* – they have both ordinary physical properties and certain non-physical properties which are not reducible to (explicable in terms of) their physical properties.

Understanding in detail the metaphysics of irreducibly non-physical properties requires a modicum of philosophical sophistication. The interested reader is encouraged to follow the suggestions for further reading to direct their research.

Again, it is likely to be less than clear why one might be inclined to maintain anomalous monism. The same material which will hopefully shed some light on the intuitions underlying epiphenomenalism should also go some way towards making clear the motivations behind anomalous monism.

CHAPTER 3

BEHAVIOURISM

The next theory of mind we're going to examine is philosophical behaviourism. Before we do so, however, it will serve us to take a short detour into the history of psychology.

One good reason for taking this detour is that the prevailing intellectual climate in which philosophical behaviourism was first formulated was one in which empirical psychology was still finding its feet as a 'science of mind'. Psychologists and philosophers were still very much trying to work out what psychology was in the business of doing and there was a concerted effort to formulate a robust philosophical theory of mind in which mentality was amenable to empirical investigation. Understanding the contemporaneous presence of psychology on the intellectual world stage gives us significant insight into the motivations of philosophical behaviourists.

Another good reason for the detour is that the term 'behaviourism' means something rather different in the mouths of psychologists than it does in the mouths of philosophers. Since there is scope for confusion here, it pays to be rigorous in disambiguating the two senses of 'behaviourism'.

Psychology is by far the youngest fully-fledged academic discipline, as it was the most recent of the disciplines to split from philosophy. It was only in the early twentieth century that psychology broke away and became an academic speciality in its own right. As such, we needn't go far back in history to trace the genesis and nascency of psychology. The story begins in Germany in the nineteenth century.

3.1 EARLY EMPIRICAL PSYCHOLOGY

The treatment of the history of psychology here will be rather cursory. For our purposes, it serves to identify a few key figures and seminal contributions which led to the birth of psychology as a distinct academic speciality. Roughly and broadly speaking, we can divide the infancy and early childhood of psychology into three stages, distinguished by the methodologies employed by the discipline's progenitors.

3.2 PHYSIOLOGICAL PSYCHOLOGY

Although we will reserve the title of 'founder of psychology' for another, Gustav Fechner (1801–87) must be credited with the inception of the empirical tradition in psychology and the delivery of the first quantitative psychological law.

Before Fechner, there was a long-standing tradition of empirical physiology, but mentality had only ever been investigated a priori, never experimentally. Fechner was trained initially as a physiologist, before becoming Professor of Physics and, later, Professor of Philosophy at Leipzig.

Fechner discovered that the way we *perceive* the intensity of sensory input is logarithmically proportional to the absolute magnitude of the stimulus. For instance, the way we perceive loudness is logarithmically proportional to the absolute magnitude of the sound waves. You may have noticed that the decibel scale which quantifies the loudness of sound is a logarithmic scale. Fechner's result has proven to be a robustly manifest quantitative relationship across the sensory modalities.

We shouldn't underestimate the importance of this result in demonstrating the possibility of, and originating a psychophysical methodology for, an empirical science of the mind. For the first time in intellectual history, we see the identification of an observable and measurable relationship between physical phenomena and mental phenomena.

The other major figure in the physiological tradition of early psychology is Hermann Helmholtz (1821–94). Like Fechner, Helmholtz had wide-ranging academic interests. He was initially educated in philosophy and philology at Potsdam and in medicine in Berlin. During his academic career he held chairs in physiology at Königsberg, anatomy and physiology at Bonn, physiology at Heidelberg and physics at Berlin. He also presided over the development of a new Institute for Physiology at Heidelberg and a new Institute for Physics at Berlin.

As well as making significant contributions to physiological optics – including the invention of the ophthalmoscope and the ophthalmometer – and delivering important unifying results in theoretical physics, Helmholtz experimented on nerve conduction. His early experiments aimed to measure the time it took for neural impulses to travel in animal limbs. He later extended this research to human subjects and, in doing so, introduced a versatile experimental technique that is still widely employed in psychology today – the measurement of *stimulus–response* times.

The groundbreaking work of Fechner and Helmholtz lay the foundation on which others would build an independent science of the mind. Fechner demonstrated the possibility of employing empirical methods to investigate mentality and identified psychophysics as a fruitful domain for further experimentation. Helmholtz was the first to demonstrate that the measurement of stimulus–response times could be a fertile methodology for the fledgling psychology.

It is of significant interest that the pioneering work of these early physiological psychologists was only possible as a result of the broad academic interests and interdisciplinary training of both researchers. This is to be taken as a cautionary note to those who would specialise too narrowly, as well as a clear endorsement of the value of cross-disciplinary analysis and interdisciplinary cooperation.

3.3 INTROSPECTIONIST PSYCHOLOGY

It is Wilhelm Wundt (1832–1920) who clearly deserves the appellation 'founder of psychology'. Wundt established the first psychological laboratory, inaugurated the first journal of psychology – *Philosophische Studien* – in 1881, and founded an Institute for Experimental Psychology at Leipzig in 1894. He also wrote the first textbook in psychology and supervised legions of graduate students from around the world who would become the first generation of psychological practitioners.

Wundt's programme of psychological structuralism placed central importance on introspection as a methodological technique. The aim of the programme was to analyse consciousness in order to identify its basic elements and the laws which connect them. This was pursued through the use of carefully designed experiments in which trained observers introspected their mental states and reported their observations.

One methodological facet of crucial importance here is the training of observers. Wundt believed that only properly qualified observers could introspect with the appropriate care and attentiveness and report their observations in a pertinent fashion, suitable for analysis.

A further point of interest lies in the rigour with which experiments were designed and conducted. Wundt and his students and colleagues were instrumental in developing many of the now standard criteria for experimentation, such as the publicity of the experimental situation, the repeatability of results and the ability to hold certain variables constant while modifying others.

Another important researcher in the introspectionist tradition was Hermann Ebbinghaus (1850–1909). Ebbinghaus established a rival journal to Wundt's – *Zeitschrift für Psychologie und Physiologie der Sinersorgane* – in 1890, and established laboratories for psychological research in Berlin and Breslau. Like Wundt, he also published an influential textbook on psychology.

Whereas Wundt and his followers thought that the scope of psychology should be properly restricted to what we might call 'lower-order' mental functions, concentrating their empirical programme solely on the investigation of mental imagery, Ebbinghaus was interested in formulating experiments with which to study memory.

In order to gain some insight into the mechanisms underpinning human memory in a fashion isolated from the potentially contaminating effects of what was already known, Ebbinghaus devised a very large number of lists of nonsense syllables – consonant-vowel-consonant segments which had no meaning in the language. He then proceeded to memorise these lists and measure, with painstaking procedures, his ability to recall these syllables.

This research delivered further quantitative psychological principles, central among them being the exponential decay of memory. Ebbinghaus discovered that his ability to recall the nonsense lists would decay very quickly at first, but increasingly more slowly through time. One consequence of this exponential decay of memory for students is the crucial importance of early reinforcement. Ebbinghaus also demonstrated that while the initial memorisation of the lists was subject to rapid decay, the *rate* of decay of the ability to recall the lists slowed in proportion to the number of repetitions. Again, this finding has obvious implications for study techniques.

Putting these two results together, we see that if one wants to be able to recall material with a high rate of accuracy, one should revisit the material very soon after first presentation, then reinforce the material after increasingly longer intervals. For instance, to recall the material presented in a lecture, it is advisable to revise the material later that day, then again a couple of days later, then again a week later, then a month later, and so on. The most cursory examination of modern advice on study techniques will yield just such recommendations.

The final figure of note in the introspectionist tradition is Oswald Külpe (1862–1915). Külpe was a student of Wundt's at Leipzig and later established a rival school of psychology at Würzburg. There were two key points of dispute between the Leipzig school and the Würzburg school. The two schools disagreed on the appropriate purview of psychology and this disagreement brought with it an associated dispute concerning experimental methodology.

Where as Wundt and his students believed that the legitimate scope of psychology was properly restricted to the investigation of mental imagery alone, Külpe and his followers thought – as did Ebbinghaus – that the empirical examination of 'higher-order' cognitive functions, such as reasoning, had an important role to play in psychology. So the Leipzig school and the Würzburg school were essentially investigating distinct aspects of mentality.

The other point of contention between the Wundtians and the Külperians concerned the role of observers in introspectionist experimentation. Wundt, as we emphasised earlier, placed significant importance in the training of observers, in order that they be qualified introspectors. Külpe, on the other hand, used exclusively *untrained* observers. History has borne Külpe out in this respect: modern-day psychological experiments typically require that the subject be not only untrained in psychology, but also ignorant of (and often deceived about) the aims of the particular experiment in which they participate.

The introspectionist tradition in early psychology was unified by the belief that introspection – whether it be by trained observers (Wundt), untrained observers (Külpe) or oneself (Ebbinghaus) – was the key to investigating mentality. Although self-reporting is still used to some extent in modern psychology (typically in combination with various other methodologies), the introspectionist tradition died out in the early twentieth century. The reasons for this are several.

Firstly, it became increasingly apparent that much of mental life is simply opaque to introspection. I can't, for instance, investigate the mechanisms governing language production and comprehension purely through introspection.

Secondly, introspection is fairly unreliable, regardless of who is doing the introspecting. People are notoriously poor at identifying their own mental states; untrained observers particularly so. Trained observers, on the other hand, have a tendency to manufacture observations in accordance with their perceived expectations. Furthermore, introspection is *itself* a mental process and therefore has an effect on the mental processes which are being introspected. If, for example, you reflect on your anger, you're likely to become either resolved and thereby less angry, or increasingly heated and thereby more angry.

Finally, only the introspecting agent is privy to the direct results of their introspection. Where introspecting subjects disagree, there is no way for a third party to adjudicate observational disputes. So while the experimental situation in which the introspection occurs can satisfy the requirement for publicity, the introspective process itself cannot.

3.4 PSYCHOLOGICAL BEHAVIOURISM

Aside from the foundational work of the physiological and introspectionist psychologists, there are two further important historical preconditions which led to the emergence of psychological behaviourism.

One of these historical antecedents was the influential nineteenth-century doctrine of positivism. Positivism, as championed by Auguste Comte (1798–1857) and Ernst Mach (1838–1916), was a reaction to the speculative metaphysics and theological conjecture which was held to have infected philosophy. Proponents of positivism held that legitimate intellectual inquiry, or 'positive science', should treat exclusively of the observable. Any doctrine which posited entities or processes beyond what could be observed was labelled with the pejorative 'pseudoscience'.

The other significant influence on psychological behaviourism was the work of Ivan Pavlov (1849–1936). Pavlov was a Russian physiologist and the originator of the theory of *reflex arcs*. Pavlov held that the connection between environmental stimulus and behavioural response was to be explained in terms of these reflex arcs. No doubt you have heard of Pavlov's dog. Pavlov first showed that the environmental presence of food stimulus would cause a dog's digestive juices to flow in its stomach, even if the food never reached the stomach. He postulated the existence of an *innate reflex arc* to account for this connection.

Most famously, however, Pavlov demonstrated that reflex arcs could be *conditioned* as well as innate. In the case of Pavlov's dog, a bell was sounded whenever food was brought into the presence of the animal. Eventually, the dog would salivate upon hearing the sound of the bell alone. This behaviour, according to Pavlov, demonstrated a *conditioned reflex arc*.

Given these contemporaneous historical factors – the burgeoning of psychology, the flourishing of positivism and the development of Pavlov's theory of reflex arcs – psychology was ripe for a paradigm shift. This was brought about by the American psychologist John Watson (1878–1958).

Watson was strongly in the grip of positivism and consequently rejected the legitimacy of introspection as a psychological tool, arguing that in order to be a positive science, psychology should concern itself only with observable behaviour. In doing so, he reconceptualised psychology – which had been seen as the science of human consciousness – as the science of human behaviour.

Watson was interested in discovering the innate reflex arcs which governed human behaviour and investigating the circumstances under which reflex arcs could be conditioned in humans. To this end he engaged in experiments on babies and young children which would certainly never be approved by today's ethics committees.

To demonstrate the presence of innate reflex arcs in young babies, Watson showed that certain environmental conditions, such as sudden loud noise or a sudden loss of support (i.e. dropping them), would evoke fear behaviour in babies. Some might think these results less than remarkable.

In investigating conditioned reflex arcs, Watson and his colleague Rosalie Rayner experimented on Albert B., an eleven-month-old infant. Albert would be presented with a white rat. When he reached out in curiosity to touch the animal, the experimenters would make a loud noise by striking a steel bar close to his head, thereby evoking fear behaviour in Albert. After only seven such presentations of the rat in conjunction with the loud noise – five of which were a week later than the first two – it was found that Albert would exhibit fear behaviour on presentation of the rat alone. This conditioned response was subsequently found to be enduring and, further, transferable to similar stimuli. Albert would exhibit the same fear response on presentation of a white rabbit and, to a lesser extent, on presentation of a dog, a fur coat or a Santa Claus mask.

Following Watson, psychological behaviourism became dominant and held sway in psychology until roughly the late 1960s. During this period it became customary to carry out experimental work on rats where possible, beginning a tradition in psychology sometimes described – endearingly or pejoratively – as occupied with 'rats and stats'.

By far the most influential psychological behaviourist after Watson was another American, Burrhus Frederic (B. F.) Skinner (1904–90). Skinner was interested in determining the most effective means of conditioning reflex arcs. He invented a device, now called a 'Skinner box', in which rats could be placed. This box contained a lever which the rats were able to depress which could be set to deliver a food pellet when actuated. By varying the conditions under which the actuation of the lever would yield a food pellet, Skinner was able to modify the behaviour of the rats accordingly.

Unlike classical Pavlovian conditioning, this behavioural modification was contingent not just on the stimulus preceding the behaviour (a ringing bell, a rat together with a loud noise) but also on the environmental stimulus *following* the behaviour. Skinner called this *operant conditioning*, and developed an associated theory of how best to effect operant conditioning by controlling the positive and negative reinforcement of certain behaviours.

Skinner argued that such operant conditioning could be widely employed as a social engineering technique. He suggested that criminal behaviour was better treated with behavioural modification techniques rather than punished through incarceration, and he published a widely read novel, *Walden Two*, which outlined his utopian vision of a planned society governed by operant conditioning. As one might expect, these ideas were met with a mixture of commendation and condemnation.

By the end of the 1960s, behaviourism in psychology had waned in popularity, in favour of the newly emerging cognitive psychology. There are good reasons for the loss of faith in the behaviourist conception of psychology.

For one thing, it became clear that positivism is, simply put, a false doctrine. Modern science is frequently in the business of theorising about unobservable entities. Such treatment of unobservables does not make theoretical physics, for instance, 'pseudoscientific'.

More importantly, it became increasingly clear that many essential aspects of mentality are simply not directly connected to observable behaviour. The mechanisms which underpin memory, the capacity to draw inferences and the ability to comprehend language do not seem to be necessarily correlated with any particular kind of behaviour. As such, these cognitive functions are not amenable to investigation in the behaviourist tradition.

While it is the case that each of the three traditions of psychology we have examined are now consigned to history, it is also the case that modern psychology preserves certain techniques from each of them. Where, however, each of these traditions were distinctly methodologically homogenous, modern psychology is markedly methodologically heterogeneous. There is still a place in psychology for psychophysical research, self-reporting is still a widely used technique and there is still a role for the observation of behaviour. There is no longer any place though for the methodological monism which characterised these early periods of psychology.

3.5 PHILOSOPHICAL BEHAVIOURISM

Psychological behaviourism, as we have seen, is a *methodological* view – a doctrine concerning the way in which one should go about doing psychology. Philosophical behaviourism, in contrast, is an *analytic* view – a substantive theory of what mental states *are*. Henceforth, when I make reference simply to 'behaviourism', I will be referring to the philosophical variety.

Philosophical behaviourism was motivated in part by positivism and the associated desire to produce a theory of mind which afforded the possibility of empirical psychology. Behaviourism was also a reaction to the serious theoretical difficulties plaguing dualism. Gilbert Ryle famously accused Cartesian dualists – in *The Concept of Mind* (1949) – of postulating a 'ghost in the machine'. Behaviourists eschewed the dualist notion that mental state terms denoted events and processes occurring in some immaterial substance.

According to the behaviourist, the terms in our language which we take to denote mental states are, in fact, simply convenient locutions for referring to complex kinds of behaviour. When we say, for instance, that Jon is in love, what we *actually* mean is that Jon goes around with a silly grin on his face, is inclined to organise his time in such a way as to see more of the object of his affections, has a tendency to write bad poetry, etc. When we say that Tillie has a toothache, what we *actually* mean is that Tillie grimaces and moans, has a tendency to clutch her jaw, is inclined to seek a dentist, etc. Attributions of mental states are, according to the behaviourist, actually attributions of kinds of behaviour: the *meaning* of mental state terms is properly specified in terms of behaviour.

Let's quickly make clear here the relation between philosophical (analytic) behaviourism and psychological (methodological) behaviourism. Clearly the former entails the latter. If mental states *just are* kinds of behaviour then the investigation of mentality is *ipso facto* the investigation of kinds of behaviour. The converse, however, does not hold. It is not the case that a methodological behaviourist must be an analytic behaviourist. One might think that psychology should be a science of human behaviour but still maintain that mental states have an existence independent of (but investigable through observing) behaviour.

Philosophical behaviourism, properly construed, is a reductive semantic thesis, according to which the analysis of mental state terms involves a *reduction* to talk of behaviour. It is crucial that this talk of behaviour is fleshed out in terms of *dispositions* to behave, not simply in terms of behaviour itself. After all, Tillie might have a toothache yet manifest no associated behaviour at all – perhaps she is fearful of dentistry and seeks to hide her pain lest she be coerced into seeing a dentist. The behaviourist accommodates this by giving essentially dispositional analyses of mental states since dispositions can be *inhibited*. In the example case, the behaviourist will argue that Tillie's toothache *just is* the disposition to groan and grimace, the inclination to clutch her jaw, the tendency to seek a dentist, etc.; however, her fear of dentistry inhibits these dispositions and accounts for her lack of manifest behaviour.

Behaviourism certainly enjoys some philosophical advantages. It solves the problem of other minds for a start, since it renounces the commitment to inaccessible immaterial mental substance. It satisfies Ockham's razor in not expanding our ontology beyond explanatory necessity, since mental states are no longer held to have any independent existence. It accommodates the notion that mentality comes by degrees, since the capacity for more complex behaviour *just is* the capacity for more complex mental states. It can account for the importance of the brain in mental life, given the nervous system is the physical cause of behaviour (hence of mentality). Finally, it confers a clear methodology for psychological investigation, in stark contrast to dualism.

Unfortunately, despite these attractive theoretical advantages, behaviourism faces a number of insuperable objections.

3.6 OBJECTIONS TO PHILOSOPHICAL BEHAVIOURISM

We'll begin with the three objections which merely problematise behaviourism and move on to the further three objections which are insurmountable for the behaviourist.

The first thing to note is that dispositional criteria can be satisfied in the absence of the associated mental state. Consider the case of actors. An actor playing a character with a toothache is disposed to moan, grimace, clutch her jaw and so on, yet is clearly not *actually* suffering from toothache.

Furthermore, dispositional criteria can fail to be satisfied in the presence of the associated mental state. Consider the case of stoics. A stoic person may have a dreadful toothache yet not be disposed to engage in any associated behaviour. The account of the inhibitability of dispositions is intended to go some way towards answering this latter objection; however, even so, the behaviourist has some difficult explaining to do. Consideration of actors and stoics and the corresponding ways in which dispositions and mental states can come apart problematises an analysis which seeks to *identify* mental states with dispositions to behave.

A third, and more troubling, objection to behaviourism focuses on how, precisely, we are supposed to *completely* specify the lists of dispositions which are identified with particular mental states. In short, the concern is that there is no way to remove the 'etc.' or the 'and so on' from the end of the enumeration of dispositions. Behaviourist paraphrases of mental state terms seem essentially incompletable, since there are a very large number of ways in which a mental state might manifest in behaviour.

Consider, for example, the statement 'Nicole loves her children'. What associated dispositions should we load into the paraphrased analysis? That she is disposed to ensure they're adequately fed, to ensure they're warmly clothed, to embrace them, to tell them that they're loved, to educate them, to be concerned for their welfare, to attend to their upsets, to comfort them when scared, to protect them from harm, to lie awake at nights thinking of their future, etcetera, etcetera.

These first three objections are certainly of concern for the behaviourist but they might be answerable with some fancy footwork. These next three objections, however, are sufficient to defeat even the most nimble-footed behaviourist.

For starters, pain *hurts*. Being in pain essentially involves a privileged first-person qualitative experience of *hurtfulness*. It is precisely this hurtfulness that characterises what it is to be in pain and distinguishes *real* pain behaviour from *pretend* pain behaviour. There is, in short, something that it is *like* to be in pain. So it is with other mental states. There is something that it *feels like* to be in love, or to be angry, or to be excited about an upcoming holiday. Behaviourism completely fails to capture these essential subjective qualitative aspects of mental states.

For seconds, dispositional analyses serve poorly as explanations. When we attribute mental states to others, we typically do so in order to explain and understand their behaviour. If mental states, however, are to be analysed in terms of dispositions to behave, these attributions are circular and uninformative.

If, for instance, I ask a physicist why it is that glass shatters when struck sharply and the reply is that this is because glass is brittle, the brittleness of glass is intended as an explanation for this behaviour. If I then inquire into the meaning of brittleness and am told that brittleness is that property that glass has such that it is disposed to shatter when struck sharply, I have learned nothing. What I was looking for was some explanation of brittleness in terms of the physical properties of the substance, not its dispositional properties.

Similarly, suppose I observe Wayne walking into a pizzeria and ask Eloise what he is doing and the response is that Wayne is hungry. Wayne's hunger is intended as an explanation for his behaviour. If I then inquire into the meaning of hunger and am told that hunger *just is* the disposition to (*inter alia*) engage in food-seeking behaviour, then I have learned nothing.

The final objection to behaviourism takes the form of an internal critique: we show that behaviourism fails by its own lights.

Behaviourism is a *reductive* theory. The whole point of the theory is to take talk of mental states and replace it, in careful analyses, with theoretical terms which satisfy positivist criteria of observability and public verifiability. The aim is to *eliminate* reference to mental states entirely by reducing talk of mental states to talk of dispositions to behave.

Behaviourist paraphrases of mental state terms, however, turn out to contain *ineliminable* reference to the mental. To say that Tillie clutches her jaw is to say more than just that her arm raises in a jaw-ward direction. To say that she seeks a dentist is to say more than just that she is impelled dentist-ward. Rather, the attributions of 'clutching' and 'seeking' are *agentive* attributions. To say these things is to say that Tillie actively, agentively and *intentionally* clutches, seeks and so on. This is already a mental attribution.

Human behaviour is always already a mental phenomenon. It is impossible to enumerate convincing dispositional paraphrases for mental terms which do not make reference to just such agentive verbs as 'clutching', 'seeking', 'organising', 'ensuring', 'attending' and so on. As such, talk of the mental is ineliminable and the behaviourist has failed to analyse the ghost out of the machine.

CHAPTER 4

NEUROANATOMY

We're now going to take a brief diversion from our examination of philosophical theories of mind and develop a rudimentary understanding of functional neuroanatomy.

The introduction to neuroanatomy here is going to be very cursory indeed. My aims in this chapter are quite modest. In the first instance, I want to show how parts of the brain are specialised for processing certain functions. In particular, we will see that our linguistic capacity is strongly localised and subserved by a rather extraordinary neurobiological adaptation.

In the second instance, I want to give a basic understanding of the operations of *neurons*. This will serve us well much later in the book when we examine artificial neural networks. Overall, I want to give a sense of just what an amazing and startlingly complex object the human brain is.

We'll begin by describing macro-neuroanatomy – the parts of the brain which can be seen with the naked eye – and then move on to describe some basic micro-neuroanatomy.

4.1 MACRO-NEUROANATOMY

The human central nervous system can be broadly divided into three areas. The spinal cord, the brain stem and the rest of the brain, including the *cerebral hemispheres* which constitute the *cerebrum*.

The spinal cord (*medulla spinalis*) is of least interest to us. It carries signals between the brain proper and the organs and muscles. Continuous with the top of the spinal cord is the brain stem which can also be divided into three parts.

The lower brain stem, or *hindbrain*, contains the *pons*, the *medulla oblongata* and the *cerebellum* (not to be confused with the *cerebrum*). The *medulla* is known to be implicated in the regulation of heart function and respiration. The *pons* (bridge) mostly relays information

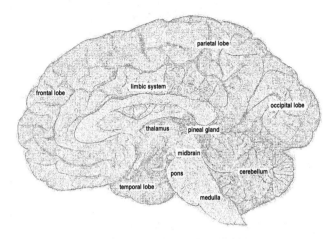

Figure 4.1   Mid-sagittal section.

between the cerebral hemispheres and the cerebellum, but is also implicated in regulating vestibular function (balance).

The human *cerebellum* (little brain) is highly distinctive. It is very densely packed with neurons – much more so than the rest of the brain – and quite regular in organisation for a neural structure of its size. The cerebellum is readily recognisable by the very fine folding of its surface, which allows for more surface area and gives it a distinctive wrinkly appearance. The human cerebellum is unique among mammalian brains in its complexity and intricacy of folding.

The cerebellum is connected to most primary sensory processing areas and most motor neurons and is known to be implicated in the automatic governing of fine motor control. When, for instance, you learn to type without thinking about it, or to operate a motor vehicle without thinking about it, your cerebellum has been programmed for the execution of a sequence of fine-grained motor responses to various sensory inputs.

The next part of the brain stem is the *midbrain*. The midbrain connects the pons to the upper brain stem. It is known to be implicated in secondary processing involved with vision and audition. It also contains the *substantia nigra* which stimulate production of the neurotransmitter *dopamine* and which play a role in assisting fine motor control. Parkinson's disease, whose sufferers experience uncontrollable fine tremors, is a degenerative condition of the substantia nigra.

The final part of the brain stem – the *upper brain stem* – contains the *thalamus* and the *hypothalamus*, as well as the *pineal gland* and the *pituitary gland*. The *thalamus*, which is continuous with the midbrain,

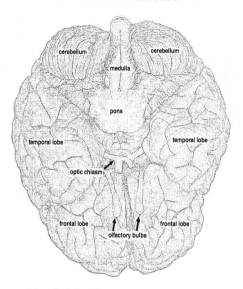

Figure 4.2 Intact brain – base view.

is often described as a 'sensory gateway'. It routes sensory information (with the exception of olfaction) from the sensory apparatus to the relevant primary processing area. It contains, among other structures, the *lateral geniculate nucleus* which is specialised for receiving information from the eyes and routing it to the primary visual processing cortex, and the *medial geniculate nucleus* which is specialised for receiving information from the ears and routing it to the primary auditory processing cortex.

The *pineal gland* and the *pituitary gland* are both *endocrine* glands – they are involved in the secretion of neurotransmitters. The pituitary gland regulates the release of a number of various neurotransmitters, including human growth hormone and *oxytocin*. The pineal gland is mainly involved with the production of *melatonin*. Interestingly, the pineal gland was believed by Descartes to be the locus of interaction with the immaterial soul.

The *hypothalamus* controls the secretions of most of the *endocrine* glands and thereby regulates a wide range of bodily functions, including thirst and appetite, temperature, and sexual and circadian cycles.

This exhausts the structures of the brain stem. The rest of the brain consists of the *cerebral hemispheres*, the *limbic system*, the *basal ganglia* and the *olfactory bulbs*.

The *olfactory bulbs* are responsible for primary olfactory processing. As mentioned earlier olfaction, or smell, is the only sensory modality which is not routed through the thalamus. Interestingly, it is also the only sensory modality which is not cross-lateralised in the hemispheres, which means that – unlike the other senses – information from the left nostril goes to the left olfactory bulb and information from the right nostril goes to the right olfactory bulb.

The *basal ganglia* are known to also be implicated in motor processing. They are closely linked with the cerebellum and the midbrain. Sufferers of degenerative conditions associated with the basal ganglia, such as Huntington's chorea, experience uncontrollable bodily movements, typically more pronounced than the tremors experienced by sufferers of Parkinson's disease.

The *limbic system*, sometimes referred to as the *limbic lobe*, is nestled around the top of the brain stem. It contains, among other structures, the *hippocampus* which is known to be implicated in memory processing and the *amygdala* which is known to play a role with respect to emotion. The limbic system is sometimes described as being part of the cerebral hemispheres.

The *cerebral hemispheres* constituting the *cerebrum* are what most people imagine when they think of the brain. It is worth mentioning that all the neural structures of the brain, not just the cerebral hemispheres, are mirrored along a line drawn longitudinally through the centre of the spinal cord. The cerebral hemispheres are joined to each other by a thick bundle of neural fibres called the *corpus callosum*.

The two cerebral hemispheres are each involved with the processing of one half of the body's sensory and motor functions and these functions are, with the exception of olfaction, cross-lateralised in the

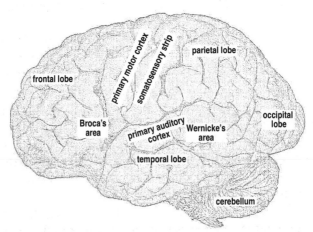

Figure 4.3   Intact brain – left view.

hemispheres. So the left hemisphere processes information from the right-hand sense organs and controls movement on the right-hand side of the body. Vision is even more peculiarly cross-lateralised. The right-hand side of each eye's visual field is processed in the left-hemisphere and the left-hand side of each eye's visual field is processed in the right hemisphere.

Each cerebral hemisphere divides into four lobes: the *frontal lobe*, the *parietal lobe*, the *temporal lobe* and the *occipital lobe*. The functions of the cerebrum are not as well understood as the more evolutionarily antecedent areas of the brain we've examined so far; however, there are certain canonical cognitive functions associated with each lobe and there are important areas of localised functional specialisation – notably the primary processing areas for each sensory modality and the two speech areas.

The *frontal lobe* is implicated in a whole range of 'higher' cognitive functions. This is sometimes loosely glossed as being responsible for 'planning and prediction' or 'executive control'. The frontal lobe contains the *primary motor cortex* which, as the name suggests, is implicated in the execution of bodily movement. It also contains one of the speech areas of the brain – *Broca's area*.

Broca's area, named after Paul Broca, is adjacent to the primary motor cortex. Damage to this part of the brain gives rise to a distinctive kind of *aphasia* (language deficit) known as *Broca's aphasia*. Broca's aphasia is characterised by an inability to produce fluent grammatical utterances, even though the sufferer retains linguistic comprehension and is aware of their deficit. This can be a particularly pernicious aphasia as sufferers struggle to make themselves understood but can only produce utterances with few, if any, grammatical particles and which are punctuated with pauses.

The *parietal lobe* is thought to play a role in integrating information from the sensory modalities. It contains the *primary sensory cortex*, otherwise known as the *somatosensory strip*. The somatosensory strip, like the adjacent motor strip in the frontal lobe, is *topologically* organised. This means that certain parts of the sensory cortex are correlated with certain parts of the body, with larger parts of the cortex devoted to those parts of the body which are more sensitive (have more nerve endings). So large parts of the somatosensory strip are devoted to the lips, fingers and genitalia, but comparatively little is devoted to less sensitive areas.

The *temporal lobe* is thought to be implicated in certain memory functions. It contains the *primary auditory cortex* and, immediately adjacent, the other speech area of the brain – *Wernicke's area*.

Wernicke's area – named after Karl Wernicke – also gives rise to a distinctive aphasia when damaged – *Wernicke's aphasia*. Wernicke's aphasia is characterised by fluent but meaningless speech. Sufferers typically evidence poor linguistic comprehension and little awareness of their deficit.

The extraordinary biological adaptation subserving our linguistic capacity which I mentioned earlier is the *arcuate fasciculus*. The arcuate fasciculus is a thick strand of neural fibres which connects Broca's area directly to Wernicke's area. Damage to the arcuate fasciculus gives rise to *conduction aphasia*, one of the distinctive symptoms of which is difficulty with repeating an utterance back to an interlocutor.

Finally, the *occipital lobe* is mostly involved with the processing of visual information. It contains the *primary visual cortex* at the very rear of the brain, which accounts for why a blow to the back of the head can cause one to 'see stars'. Damage to the occipital lobe can result in blindness, even when the visual sensory apparatus remain intact and functional.

Although there is considerable localisation of function in the brain, it does exhibit a certain degree of *neural plasticity*, particularly in younger brains. This means that if a certain part of the brain is damaged, other parts of the brain may be able to take up its function to some extent. This is generally more so with the functions implemented in the cerebral hemispheres. Damage to the brain stem is usually irreversible and will typically quickly lead to death since these areas regulate vital functions.

4.2 MICRO-NEUROANATOMY

The final aim of this chapter is to briefly describe the operations of *neurons*. Neurons are individual nerve cells which conduct electrical impulses and the brain consists of a very large number of them.

There are roughly ten billion neurons in the brain, each of which is connected, on average, to about ten thousand other neurons. That makes brains *astonishingly* complex. Imagine taking a country the size of India – which has a population of about a billion – and giving every man, woman and child a thousand pieces of string with instructions to find a thousand distinct people to hold the other end of each piece of string. When the whole country is connected up like this, with every person connected to a thousand other people by pieces of string, multiply the whole system in complexity by an order of magnitude and *that's* how complex your brain is.

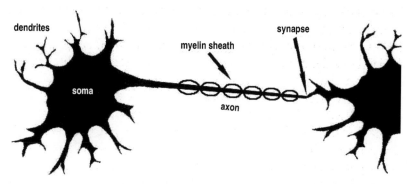

Figure 4.4 Model neuron.

There are a number of quite distinct types of neurons, but that needn't concern us here. We're going to describe the structural features and operations of a paradigm neuron.

Neurons have a *cell body* or *soma* which contains the *nucleus* of the cell. This is connected via an *axon hillock* to the *axon*, which is a protuberance which can extend as long as roughly a metre. These axons are coated with a *myelin sheath* which helps electrical signals flow more quickly and aids in insulation. At the end of the axon are *axon branches* which terminate at *axon terminals*.

Axons are *efferent* connections – they carry signals away from the soma and along the axon towards the axon terminals. Incoming, or *afferent*, signals are carried towards the soma along the *dendrites* of the neuron. Dendrites are organised in a *dendritic tree* and there may be very many of them.

When an axon terminal is in close proximity to a dendrite, a *synapse* will form (see Figure 4.5). These *synaptic connections* conduct signals from one neuron to another (strictly speaking, a neuron can form a synaptic connection with pretty much any part of a neighbouring neuron, but we're aiming to keep things as simple as possible).

The operations of neurons are electrochemical in nature. An electrical signal flows along an axon to a *presynaptic* axon terminal where it is *transduced* into a chemical signal. This chemical signal is then carried across the *synaptic cleft* by neurotransmitters in *synaptic vesicles*.

Once these synaptic vesicles reach the *postsynaptic* structure, the chemical signal encoded by the neurotransmitters is *transduced* back to an electrical signal which propagates along the postsynaptic dendrite and into the body of its neuron.

The cell body of a neuron has a certain electrical *resting potential*. As it receives afferent electrical signals along its dendrites, the

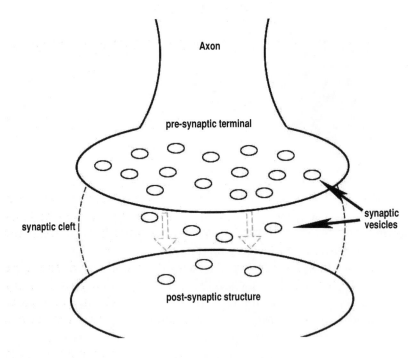

Figure 4.5   Model synapse.

difference in electrical potential between the inside of the cell and the outside of the cell rises. When this potential difference is high enough, the soma will discharge an electrical impulse along its axon and return to its resting potential. This is something of a simplification but it suffices for our purposes.

If you're feeling a little overwhelmed with all these technicalities and all this new terminology, don't fret. All you really need to take away from this chapter with respect to the operations of neurons is the following.

There are very many neurons in the brain which are highly interconnected. These neurons function by passing electrical signals to each other. If a neuron receives sufficient incoming signals from other neurons, it will send out a signal of its own.

Those readers who are intrigued by what they've read here and would like to learn more are advised to follow the suggestions for further reading.

CHAPTER 5

AUSTRALIAN MATERIALISM

Now that we have at least a rudimentary understanding of just what an amazing thing the human brain is, it is time to examine a philosophical theory which posits a very strong connection between the neural and the mental.

Australian materialism – so called as its major proponents were located in Australian universities – is a theory which goes by many names. It is variously also known as reductive materialism, identity theory, type physicalism and central state materialism, for reasons which will become apparent in due course.

It will serve our purposes here to develop Australian materialism in conjunction with another theory: the causal theory of mind.

When we ask the question 'what *are* mental states?', there are two distinct kinds of answer one can provide. One kind of answer involves giving a conceptual analysis of mental states – an account of what we *mean* by 'mental states'. Another kind of answer involves providing a substantive identification – indicating *which things* turn out to be mental states.

In the case of behaviourism, these two kinds of answer to the question of what mental states are were conflated in the one theory. This is because behaviourists are *eliminativists* about mental states and, hence, do not believe there is a substantive identification to be made. They hold that talk of mental states is, in fact, just talk about dispositions to behave – behaviourism is an ontologically eliminative and semantically reductive theory of mind.

We're now going to employ more philosophical sophistication and carefully tease apart the two ways of answering the question of what mental states are. Australian materialism will provide us with the substantive identification – the account of *which things* turn out to be mental states. The account of *what it is to be* a mental state, however, will be provided by the causal theory of mind.

5.1 THE CAUSAL THEORY OF MIND

The canonical exposition of the causal theory is given by David Armstrong in his 1968 monograph, *A Materialist Theory of Mind*. Armstrong, together with J. J. C. Smart and U. T. Place, is one of the three major figures associated with Australian materialism.

The causal theory, as we have said, aims to give an account of *what it is to be* a mental state. This is, if you will, very much like providing a job description for mental states. A job description does not specify the race, age or gender of the occupant of the role. It merely tells us what the relevant duties are – what one has to *do* in order to fill the role. So it is with the causal theory – it tells us what something has to *do* in order to fill the role of a mental state.

We begin by reflecting on the fact that many terms in our language are defined by reference to their causal powers. The term 'poison' is a paradigm example. A poison can be a liquid, a solid or a gas. Poisons can be coloured or colourless; they can be odourless or have a distinctive odour, and so on. None of these properties are relevant to whether or not the substance in question is properly called 'poison'. What makes a substance a poison is its causal role with respect to bringing about ill health in humans. To cast this as a definition we can say the following: a substance is a poison iff it is apt to cause ill health in humans.

The central tenet of the causal theory is that mental state terms are just such terms. We define mental state terms, according to the causal theorist, by reference to their causal role with respect to behaviour.

This allows the causal theorist to provide a schema for defining particular mental states along the following lines. To be in a particular mental state is to be in a state which is *apt to be caused* by certain stimuli and *apt to cause* certain behaviour.

Providing a definitional taxonomy of mental states is then simply a matter of identifying the stimuli which are apt to cause each particular mental state and the behaviour each is apt to cause. To be in pain, for instance, is to be in a state which is apt to be caused by, *inter alia*, burning my hand and is apt to cause, *inter alia*, the removal of my hand from the source of heat and further hand-discriminating behaviour.

The causal theory conserves, from its philosophical predecessor, the intuition that there is a crucial connection between mentality and behaviour. It does not, however, make the problematic identification between mental states and dispositions to behave. To be in a mental state is not, for the causal theorist, just to be disposed towards certain behaviour. Rather, to be in a mental state is *to be in a state* which is apt to stand in certain causal relations mediating stimulus and behaviour. The causal theorist is not an eliminativist about mental states.

The obvious next point of inquiry is to determine precisely which states are apt to have these causal properties. In other words, now that we have characterised the *role* of mental states by giving a conceptual analysis of mental state terms, it is time to locate the *occupants* of these roles by making substantive identifications.

The causal theory, as it stands, is ontologically neutral – it does not commit us to any particular ontology. One could, for instance, be a causal theorist but still maintain that the things which occupy the roles of mental states are immaterial. This would, of course, require an unusual account of causality but we have already seen that this is a problem for the dualist.

Although the causal theory is, strictly speaking, ontologically neutral, the talk of causality does pave a fairly obvious path to a material identification of mental states. This is precisely what is provided by Australian materialism.

5.2 THE IDENTITY THEORY

Australian materialism rose to prominence in the late 1950s with the publication of two very influential papers: Place's *Is Consciousness a Brain Process?* (1956) and Smart's *Sensations and Brain Processes* (1959).

Australian materialism makes strict identifications between types of mental states and types of neural states. In other words, to be in a certain type of mental state *just is* to be in a particular type of neural state.

This is an analysis of mental states that aims to provide an *intertheoretic reduction*. Types of mental states, according to the Australian materialists, smoothly reduce to types of neural states. They do not hold to eliminativism with respect to mental states but, rather, seek to make scientific identifications of the correct referents of our mental state terms.

So the Australian materialist does not believe, as did the behaviourist, that we are simply mistaken in using mental state terms as if they refer to substantive entities. Where the behaviourist held mental state terms to be akin to terms like 'witches' and 'phlogiston' – to be shown to be mere 'folk' terms by the progress of scientific discovery – the Australian materialist holds that mental state terms are akin to terms like 'lightning', to be identified with physical phenomena in accordance with our scientific theories.

At this point, the various names by which the theory is known should make a lot more sense. It is a reductive materialist theory that makes type–type identifications between mental states and certain physical states – namely, neural states, or states of the central nervous system – hence the appellations 'reductive materialism', 'identity theory', 'type physicalism' and 'central state materialism'.

One clear advantage of the theory is that it provides a solution to the problem of other minds. We can tell whether other people actually have mental states simply by investigating their brains. Having a type of neural state, on this analysis, *just is* having a type of mental state, so other minds are readily identifiable and empirically amenable.

Another clear advantage of the theory is that it confers a scientific methodology for investigating mentality. If we want to know about the mind, we should do neuroscience. In particular, we should seek to determine which types of neural states obtain as which types of mental states.

As a materialist – or physicalist – theory, Australian materialism also satisfies Ockham's razor. The Australian materialist admits only physical substance into her ontology. At least, *qua* Australian materialist, this is the case – she may well have other reasons to expand her ontology, but these won't be reasons which pertain to her theory of mind.

A final selling point lies in the theoretical fit with the causal theory of mind. To the extent that one holds that the causal theory is a correct analysis of mental states, one finds an advantage in the provision, by Australian materialism, of candidates that are apt to have precisely the causal powers held to be characteristically defining of mental states.

If we, then, marry the conceptual analysis of the causal theory with the substantive identification of Australian materialism, we get the following account of mental states. To be in a type of mental state is to be in a type of neural state which is apt to be caused by certain stimuli and apt to cause certain behaviour.

Despite the numerous advantages of Australian materialism, there are, as always, a number of philosophical objections we can mount against the theory.

5.3 ARGUMENTS AGAINST AUSTRALIAN MATERIALISM

Let's begin with some fairly weak objections to Australian materialism. We might argue that we have the capacity to introspect our mental states and that when we do so, we learn about our mental states. We don't, however, learn anything about our neurophysiology through introspection, so mental states can't be identical to neural states.

There is a clear reply to this objection. It straightforwardly *begs the question* against the Australian materialist. 'Begs the question' is a phrase which is more and more commonly used to mean something along the lines of 'in light of which the question demands to be asked'. This is *not* the technical philosophical use of 'begs the question'. To beg the question is to run afoul of the fallacy of *petitio principii*. One begs the question against one's interlocutor when one asks to be granted the very proposition in dispute.

In this case, the proposition in dispute is that mental states are type-identical to neural states. The objection from introspection only cuts any philosophical ice on the assumption that mental states are *not* identical to neural states. After all, if the Australian materialist is correct and this type identity holds, we *do* in fact learn something about our neural states through introspection, simply by virtue of learning about our mental states (given these are held to be identical). The objection from introspection therefore simply begs the question and is no real objection. It may be surprising to learn that we do actually introspect our neural states but scientific discovery is frequently surprising in the light of antecedent folk theories.

There are a number of further objections to Australian materialism which also beg the question. Several of these come in the form of appeals to Leibniz' Law – the objection from introspection is actually an instance of just this form.

Leibniz' Law – otherwise known as the *identity of indiscernibles* – posits that if two things have all and only the same properties, then they are identical. The objection from introspection seeks to deploy this in arguing that since mental states have a property which neural states do not, they must be not identical. Similar objections can be mounted by appealing to various other properties the objector holds mental states to have and neural states to lack, or vice versa.

For instance, neural states have a specific spatio-temporal location. It seems odd, however, to suppose that my mental state of thinking about ice cream is located three inches behind my right eye.

Alternatively, we might play on the semantic properties of mental states. My mental state of believing that today is Saturday has *semantic content* – it *means* something. By virtue of its semantic content it is apt to be involved in implication relations – for instance, *if* I believe today is Saturday *then* I believe tomorrow is Sunday. Neural states, however, neither have semantic contents, nor are they apt to be involved in implication relations.

There are a number of other ways we could problematise the appeals to Leibniz' Law here; however, the most decisive reply is to simply note that, once again, these objections straightforwardly beg the question against the Australian materialist. If mental states and neural states are, in fact, type-identical, then mental states *do* have a specific spatio-temporal location – surprising as this may be – and neural states *do* in fact have semantic contents such that they are apt to be involved in implication relations, surprising as this may be. To simply *assert* that this is not the case is just begging the question.

There is, however, a rather decisive objection to Australian materialism. This is the objection from multiple realisability.

It turns out not to be terribly difficult to cast serious philosophical concerns over the claim that types of mental states are identical to types of neural states. For one thing, there is the question of how, precisely, we are supposed to construe the concept of 'type'. If we construe it too narrowly then we are committed to saying that whenever a group of us all desire ice cream, we are each in *exactly the same* neural state. This is clearly implausible and the rubric of 'type' is supposed to allow for some variation in neural states in order to accommodate this. How *much* variation is the crucial question. If we construe the notion of 'type' too broadly, then we are at risk of losing the empirical methodological advantage which the claim of type-identity confers.

The explanatory burden here on the Australian materialist is to give some account of what, precisely, must be shared by neural states in order for them to qualify as being of the same type. Unfortunately for the Australian materialist, even if this explanatory burden can be met, there is a further objection to the type-identity claim which is unanswerable.

Consider the case of someone who suffers neural damage, whether it be through a stroke or through some trauma such as a motor vehicle accident. After the damage, the patient typically loses the ability to have certain mental states. They might, for instance, no longer be able to recognise their spouse, or they might no longer be able to understand certain words. The fact of crucial importance here is that such patients very frequently regain many of their lost mental faculties – they *relearn* to recognise their spouse or to understand the concepts they had lost. They recover the capacity to have these mental states *despite* the fact that the neural substrate which originally supported these functions is irrevocably damaged.

Before the damage occurred, being in mental state x meant being in neural state x. After relearning the lost mental functions, however, being in mental state *x* means being in a *totally distinct* neural state *y*. This neural plasticity – the ability of parts of the brain to take up functions that are ordinarily carried out by quite distinct parts of the brain – is well documented. The demonstrable multiple realisability of mental states provides a decisive refutation of the type identity posited by the Australian materialist.

Mental states then are not only multiply realisable across subjects in such a way that problematises the rubric of 'type', they are also demonstrably multiply realisable *in the same subject* in such a way that refutes the claim that types of mental states are identical to types of neural states.

Furthermore, it is worth briefly noting here that Australian materialism is prejudiced against the possibility of non-human minds. If mental states are held to be type identical to states of the human central nervous system, then it is not possible for dogs and cats, for instance, to have mental states. I'm quite certain though that our cat Linus and our dog Mia have mental states. They certainly don't have the complexity of mental states or cognitive powers that humans have – far from it – but it seems implausible in the extreme to argue that they lack beliefs and desires.

Australian materialism rules, by fiat, against the possibility of mental states obtaining in non-human biological substrates. Further – and crucially for our purposes – it rules against the possibility of artificial intelligence. This is not, in and of itself, much of an objection but it is certainly what we might consider an untoward – and unmotivated – consequence of the theory.

In light of the multiple realisability objection, it is clear that one cannot continue to maintain type–type identity between mental states and neural states. One possible modification of the theory is to retreat to a type-token identity. This is simply to argue that whenever one is in a particular type of mental state, there is an associated token neural state. In other words, to be in a type of mental state just is to be in *some* neural state. Mental states are still taken to be identical to neural states, but no particular type of neural state is held to be a particular type of mental state.

While this modification accommodates the multiple realisability of mental states – both across subjects and within the same subject across time – it makes for a very weak theory indeed. The theorist who posits type-token identity is no longer making the kind of identification that facilitates an intertheoretic reduction. We can no longer investigate mentality by doing neuroscience since there is no advantage in determining which mental states obtain as which neural states.

If the identity is only a token identity, then this determination will only hold for the subject under investigation at the time of investigation. The results are not universalisable in the way they were for the theorist positing type identity.

The token identity theory then – or token physicalism as it is sometimes known – is barely worth entertaining. One of our desiderata for the philosophical adequacy of a theory of mind is its empirical adequacy – ideally our theory of mind should direct empirical investigation.

Fortunately, we now have the makings of such a theory at our disposal and the purpose of the following chapter will be to develop it. We're going to do this by preserving the core intuition of the causal theory of mind in such a way that allows for the motivations which underpin Australian materialism, but without overcommitting in the way their substantive type–type identification does.

Before we do so, however, it will serve our purposes to examine a very well known thought experiment and consider one possible argument we might draw from it.

5.4 WHAT MARY DIDN'T KNOW

Thought experiment plays an important role in the philosophy of mind. Since this is the first time we are seeing one in this volume, it is worth very briefly discussing their role.

Thought experiments aim to prime our intuitions by asking us to imagine certain logically possible situations. By their very nature, they typically describe wildly outlandish and implausible situations and the following thought experiment is no exception. It does no philosophical work, however, to simply object to the physical possibility of the thought experiment situation obtaining, although this is a common response when first meeting them. To engage philosophically with thought experiments is to identify logical consequences of the situation being described – to argue that such-and-such *must be* the case *were* the situation to obtain.

With that in mind, let's consider the case of Mary, empirical scientist par excellence. Mary has access to completed physical theories – not just our current best theories but *completed* theories. Mary has been assiduously studying these theories for rather a long time and has reached the point where she knows *all* the physical facts. In particular, Mary knows *everything* there is to know about colour. She knows all about wavelengths of light and the reflectance of various surfaces. She also knows all about human neurophysiology so she knows all about human sensory apparatus and the visual capacity. She knows *all the physical facts there are to know* about human colour experience.

Mary, however, has never actually had a colour experience herself. Up until now, there has been a device implanted in her brain which prevents her from seeing in colour – her entire world to date has been a world of black and white experiences. As a reward, however, for finally learning everything there is to know about the physical world, this device is remotely disabled, allowing Mary the capacity to have colour experiences. The first thing Mary sees after the device is disabled is a red rose. Mary has an experience she has never had before – the experience of seeing the colour red. And she *learns something new*. She learns something she didn't know before, even though she knew all the physical facts about the world. She learns *what it is like* to see the colour red.

This thought experiment was originally intended to support an argument against *any* theory which seeks to account for mentality in purely physical terms. The argument runs along the following lines. Mary knew all the *physical* facts. Yet Mary learned something new when she had a colour experience. So there is more to know about mental life than is provided by all the physical facts. Hence, a purely physical account of mentality is not a complete account.

What is left out, it is contended, is an account of *what it is like* to be in a mental state. These privileged subjective qualitative aspects of mental states – *what it is like* to have the experience of being in a particular state – are termed *qualia*. We touched on qualia briefly in Chapter 3 when we discussed the hurtfulness of pain as an objection to behaviourism. The given argument against physicalism makes the scope and substance of that objection explicit.

It is undeniable that there is something that it is like to be in any given state that can only be known by having the first-person experience of being in the state. What is contentious is precisely what explanatory burden, if any, is conferred on theorists of mind by qualia.

The thought experiment above was originally described – albeit in a slightly different form – by the Australian philosopher Frank Jackson in his 1982 article *Epiphenomenal Qualia*. The argument against the explanatory adequacy of purely physical theories of mind we have presented here – which is also found therein – is known as the knowledge argument. There is an enormous literature surrounding this argument which I do not intend to summarise here. Once again, I refer the interested reader to the suggestions for further reading.

Rest assured, however, that this is certainly not the last we will see in this volume of the qualia issue. For the moment, though, it is time to turn our attention to an account of functionalism.

CHAPTER 6

FUNCTIONALISM

There is very little explanatory work to be done in this chapter and, consequently, it will be comparatively short. The reasons for this are twofold.

One reason is that we have done much of the required setting up for functionalism in developing the causal theory of mind. This was one of the motivations behind presenting an account of the causal theory in conjunction with Australian materialism.

The other reason is that functionalism is, strictly speaking, a *theoretical framework* which requires fleshing out into a fully-fledged theory of mind. One of the ways of fleshing out the functionalist framework gives us the theory which it is the central concern of this volume to develop and evaluate – *computationalism*. Before we fully develop computationalism, however, we are going to suspend our discussion of philosophical theories of mind and work up a rigorous technical account of precisely what *computation* is.

For the moment, we will be satisfied with making the clear the structure of the theoretical framework which we will later develop more fully.

6.1 FUNCTIONAL DEFINITION

We begin – as we did with the causal theory – by reflecting on the defining characteristics of certain terms.

Many terms in our language are defined by the characteristic function of their referents. A paradigm example here is the term 'carburettor'. A carburettor is a (now largely obsolete) device in internal combustion engines whose function is to mix fuel and air in precise ratios for maximally complete combustion. Carburettors can be made of metal alloys, or ceramic, or some other material. They can employ butterfly valves or sliders or other mechanisms to regulate their inputs and output. They can have one chamber or multiple chambers. They can be any colour one chooses. None of these characteristics, however, have any bearing on whether or not something is properly called a 'carburettor'.

Something is a carburettor iff it serves the function of a carburettor – iff it mediates between fuel and air inputs and a combustion mixture output. Anything at all which can serve this function counts as a carburettor.

The functionalist, as you have no doubt guessed by now, holds that mental state terms are precisely such terms. What makes a state a mental state is not, according to the functionalist, anything intrinsic to the state but, rather, its function in mediating relations between inputs, outputs and other mental states. Mental states are held to be *functional* states.

Understanding any particular type of mental state is, on a functionalist analysis, simply a matter of understanding its function. Pain, for instance, is held to be a functional state which in humans is characteristically caused by bodily trauma and which characteristically causes distress and reasoning aimed at alleviating the pain, as well as characteristically causing behaviour which is aimed at seeking relief from the pain.

In other words, pain is a functional state which mediates relations between characteristic pain-inducing inputs, pain-alleviating reasoning and behaviour. Anything at all which is apt to serve this function – to mediate relations in such a way – *just is* a pain state.

No doubt you are thinking this is sounding very much indeed like the causal theory. If you are, then you are certainly correct – the causal theory is, in fact, an early form of functionalism.

You may also be thinking that this is rather reminiscent of behaviourism. Once again, you would be correct. There is a sense in which functionalism is the new behaviourism. The functionalist account preserves the important connection between stimulus, mentality and behaviour. The crucial distinction, however, is that the functionalist is *not* an eliminativist about mental states. In fact, the functionalist holds that any adequate description of mental states contains an ineliminable reference to *other mental states*.

According to the functionalist, the characteristic function of mental states is to mediate relations between inputs, outputs and *other mental states*.

6.2 A BLACK BOX THEORY

In order that we fully appreciate the structure of the theoretical framework, it is useful to represent the three levels of description – and

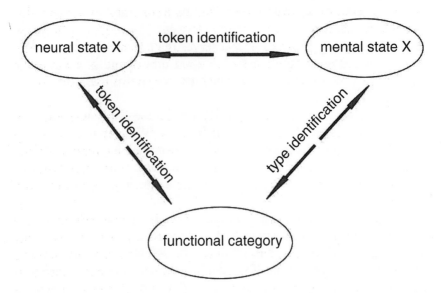

Figure 6.1   Type/token identity.

the identities that obtain among them – diagrammatically (see Figure 6.1).

As we can see, there are both type identifications and token identifications involved in the functionalist framework. A particular state – I've used a neural state for the sake of example – is identified with a particular mental state. However, unlike Australian materialism, this is not a *type* identification. The neural state is *token*-identified with the mental state by virtue of being token-identified with a particular functional role which itself is type-identified with the mental state.

In other words, types of mental states *just are* types of functional roles. So mental states are type-identical to functional categories. Any state which is apt to carry out that function *just is* the mental state but *not* by virtue of any intrinsic properties of the state. This is why the identity is only a token identity. Anything at all could stand in place of the neural state in Figure 6.1 if it carries out the appropriate function.

Yet another way of saying this is that the neural state, in Figure 6.1, *happens to be* identical to the mental state in this instance – it *happens to be* the thing which *is* the mental state – but not by virtue of being any particular *type* of neural state. Any type of neural state (in fact any type of state at all) can *happen to be* identical to the mental state if it *happens* to carry out the requisite function.

Unlike the token physicalism we briefly considered at the end of the last chapter, functionalism is not methodologically vacuous with respect to empirical investigation. Quite the contrary. By virtue of the type-identification between mental states and functional categories, we know precisely how we should investigate mentality empirically. Psychological inquiry, on the functionalist account, is a matter of determining and investigating the characteristic functions of particular types of mental states.

This is one clear advantage of the functionalist framework – it directs psychological inquiry in just the way we require from an empirically adequate theory of mind. To better understand mentality is to develop an account of the mediation by particular types of mental states of their relations to characteristic inputs, outputs and other mental states.

A further advantage of functionalism lies in its preservation of the intuitions which underwrote precursor theories. The behaviourist intuition that mentality crucially involves relations between stimulus and behaviour is preserved and the Australian materialist intuition that mental life is to be accounted for in terms of neural activity is accommodated.

It is important to appreciate, however, that while one is at liberty to identify neural states as those things which serve the role of mental states, one is not *committed* to doing so simply by virtue of committing to functionalism.

It is an important feature of functionalism that it is *substrate independent* and, hence, ontologically neutral. Anything at all – including a state of non-physical substance – can be a mental state on the functionalist account, provided it carries out the requisite function. This substrate independence is precisely what allows the functionalist to accommodate the multiple realisability of mental states without succumbing to methodological vacuousness.

A corollary of this substrate independence is the avoidance of the species chauvinism inherent in Australian materialism. As far as the functionalist is concerned it is an open question whether or not non-human entities – biological or otherwise – have mental states. Functionalism allows for the possibility of dog minds, cat minds, Martian minds and – crucially – man-made artefacts with minds. The functionalist framework allows for the possibility of artificial intelligence.

As well as enjoying substrate independence, functionalism is also mechanism independent. It says nothing about the actual mechanism by which mental states carry out their function in mediating relations between inputs, outputs and other mental states. For this reason, functionalism is often called a 'black box' theory of mentality.

Mental states, on the functionalist account, are akin to black boxes. We know neither what they are made of, nor what goes on inside them. While this confers the theoretical advantages we have described, there is also a sense in which one is left wanting by the functionalist account of mental states. One wants to know more, in particular, about the details of the mechanisms which facilitate the mediation held to be characteristic of mental states. This is why I refer to functionalism as a *theoretical framework*. Different ways of specifying the mechanism in question result in various fully-fledged functionalist theories.

A prime candidate for a mechanism which is apt to carry out precisely such mediation is *computation*. Fleshing out functionalism with a computational account of the mediating mechanism will deliver us the theory we are centrally concerned with – *computationalism*.

In order that we might do so responsibly and accurately, we are going to need a rigorous formal account of just what *computation* is. This will be the target of the next three chapters.

Before we move on to this formal material, however, let's briefly consider a couple of standard philosophical objections to the broad functionalist framework.

6.3 QUALIA OBJECTIONS

The two objections we will raise here target not any particular kind of functionalism but, rather, the claim at the heart of the functionalist framework. These are objections to the contention that there is nothing more of importance to know about mental states beyond their function and that carrying out such a function is *sufficient* for something being a mental state.

Both objections seek to highlight the importance of *qualia* in mental life and aim to establish an explanatory burden on theorists of mind to account for qualia. Let's first consider the *inverted spectrum* objection.

Whenever I'm in the presence of objects with certain surface reflectance properties, under certain lighting conditions, I have a perceptual experience such that if someone asks me what colour I perceive the object in question to be, I will respond 'blue'. The experience I have when observing the Pacific Ocean in bright sunlight is a paradigm example of what I refer to by the colour term 'blue'.

Whenever Sue is in the presence of the same objects under the same lighting conditions and I ask her what colour she perceives, she also replies 'blue'. Whenever I see an object which I perceive to be yellow and say 'that looks yellow', Sue is in agreement. So it is with all our other colour terms. In other words, we concur always and everywhere on the extension of our colour terms.

The problem is that I don't have direct access to Sue's perceptual experiences, only to her reports of her experiences. Given that we always and everywhere point to the same things when uttering colour terms, I *presume* that when Sue has a perceptual experience which she reports as 'blue' that she thereby has a perceptual experience *just like* the one I have when I experience blueness. For all I know, however, it may well be the case that whenever Sue perceives what she reports as 'blue', she is *actually* having a perceptual experience such that if *I* were to have that experience I would report it as being 'yellow'. In fact, for all I know, this may be the case for all our colour terms – our colour spectra may be completely inverted with respect to each other.

The argument against the adequacy of a functional account of mental terms which we draw from this thought experiment runs along the following lines. My mental state of 'perceiving blue' and Sue's mental state of 'perceiving blue' are *functionally equivalent*. Our respective states mediate characteristic blue-type stimulus, blue-perceiving behaviour and other mental states in just the same way. According to the functionalist then, our respective states are *equivalent* and there is nothing to distinguish the two. After all, the only possible distinction between mental states, on a functionalist account, is a distinction in function. It seems quite clear, however, that there is something quite different about our respective states of 'perceiving blue'. Sue, by virtue of her inverted colour spectrum, is perceiving what I would call 'yellow'.

It is not at all clear what to say about this argument and I certainly don't intend to rule on it here. The literature, as I have said, is very much divided on the importance of qualia.

On the one hand, it seems clear that there is an important distinction between the two mental states in question which cannot be accounted for in terms of function alone. What it is like for me to perceive blue is not at all what it is like for Sue to perceive blue. Quite distinct qualia attach to the two experiences.

On the other hand, it is not obvious just what hangs on this. After all, if we are always and everywhere in agreement with respect to the extensions of our colour terms, surely this is all that is important. Does it really matter that I would call her 'blue' experience a 'yellow' experience? Especially since this is something that neither of us, nor anyone else, could ever know?

The second objection against the adequacy of a functional account of mentality is known as the *absent qualia* objection, or – more entertainingly – the *zombie objection*.

Consider, if you will, a being indistinguishable for all intents and purposes from you and me. Let's call this being Imitation Man. Imitation Man has a regular life, just like you and me. He has likes and dislikes, goals and ambitions, beliefs and desires – in short he is just like any other human being. When asked about his experiences, Imitation Man will give the kind of answers we would expect any other person to give – he will tell us that his pain hurts, that the experience of listening to certain music is pleasurable and that chocolate ice cream tastes marvellous.

Unlike you and me, however, Imitation Man is completely lacking in qualia. He is what we might call a 'zombie'. His pain doesn't actually *feel like* anything. There is nothing that *it is like* for him to listen to music or to taste ice cream. There is no way for us to ever discover this though since we do not, of course, have direct access to his experiences.

This in no way speaks against his *functional equivalence* with other human beings. His pain state still plays the functional role that pain states play in everyone else, as do all his other mental states. They simply don't *feel like* anything.

The argument to be made against functionalism here should be obvious. Imitation Man's pain state, for instance, is functionally equivalent to ours – it mediates relations between stimulus, behaviour and other mental states in just the way our pain state does. If Imitation Man's hand is placed on a stove, he will yell, cry 'ouch' (or some other appropriate expletive), seek to remove his hand from the source of heat, engage in hand-soothing behaviour, weep and moan, and so on. Imitation Man's pain and our pain are, therefore, equivalent *simpliciter* on the functionalist account . Yet it seems clear that Imitation Man's pain is a rather different thing to our pain – his pain doesn't *hurt*.

As with the inverted spectrum objection, it is not at all clear what one should say to the absent qualia objection. Once again, I shan't rule on it here but will simply indicate the two kinds of ways one might be tempted to respond.

On the one hand, it seems that the example is unfairly prejudiced against the adequacy of functional explanation. It assumes that carrying out all the various functions characteristic of human mental life is not *ipso facto* having qualia. In fact, it is tempting to charge the absent qualia objector with begging the question against the functionalist.

However, there is nothing in the functionalist account so far which maintains the presence of qualia as a straightforward consequence of carrying out the various functions characteristic of mentality.

The functionalist only maintains that a functional account of mental states is an *adequate* account, so the absent qualia objector doesn't *quite* beg the question. It is difficult to appreciate, however, how Imitation Man's pain state could be functionally equivalent to ours without his pain being hurtful. After all it is the very hurtfulness of our pain which motivates our pain alleviating behaviour, is it not? There is still a sense in which the absent qualia objection is somehow loaded against the functionalist.

On the other hand, the absent qualia objection seems to bring out *precisely* the inadequacy of a functional account of mental states. The functionalist account of mentality fails to capture what seem to be essential aspects of our mental states – their subjective qualitative aspects. There *just is* something that it is like to see blue, or to be in pain, or to taste chocolate ice cream, and one might think that any theory which fails to give an account of such qualia is explanatorily inadequate.

With these difficult philosophical issues to ruminate on, it is time to turn our attention to a formal account of computation.

CHAPTER 7

FORMAL SYSTEMS

In the previous chapters, we have considered the question of what minds *might be* and sketched out the space of possible responses to this question. In doing so we have seen a progression of philosophical theories of mind and considered arguments and objections pertaining to each.

In the following three chapters, we are going to be working our way towards a precise formal account of what computers *are*.

Unlike the question of what minds might be – which is ripe for theorisation – there is something that it *is* to be a computer and specifying that something is a purely descriptive exercise which involves delving into theoretical computer science and teasing out some foundational material.

In these chapters, I presuppose no understanding whatsoever of computer science, mathematics or any formal discipline. If you have an aversion to symbols then do not fear. The introduction here is deliberately slow and gentle and there are numerous exercises to aid understanding.

We will start in this chapter by defining *formal systems* and playing with some toy (simple) formal systems to get a basic feel for symbol manipulation. We are then going to spend the next chapter investigating a particular kind of formal system: a *register machine*. We will use the concept of a register machine – and related concepts involved in its explication (like *program*) – to give a precise characterisation of *computability*.

With this rigorous definition of computability, we can then speak authoritatively (and correctly) about *computing*, *computers* and *computation*. We are going to play with some toy register machine programs to get a feel for the syntactic nature of computation. We will also have a look at some more difficult and complicated register machine programs (for those who are amenable to such things and enjoy a challenge). These more difficult challenge exercises can be skipped without prejudice by those who have no taste for them.

Finally, we complete our survey of computational theory in Chapter 9 by seeing how we can use the very clever method of *Gödel coding* to define a *universal machine* – a machine which can compute *any* computable function. We shall also discuss just what it is to be computable – what falls within the limits of the computable and what falls outside.

Armed with a sound knowledge of computational theory, we will have precise formal definitions and some subtle distinctions at our disposal. Deploying these, we will be able to correctly and responsibly characterise the theory that is our central concern. That will be our first aim in Chapter 10.

7.1 EFFECTIVITY

It is highly likely that every reader of this book has at some stage in their life played a game of at least one of the following: chess, draughts, backgammon, go, Chinese checkers or – at the very least – tic-tac-toe (aka noughts and crosses). If you understand how at least one of these games is played (most of us can grasp tic-tac-toe), then regardless of how good or bad you are in playing them, you *already understand* the principles underlying formal systems. We'll begin our examination of formal systems by simply making explicit what you already grasp implicitly.

Chess exemplifies the important features of formal systems nicely, so I will make reference to it throughout this chapter. Don't be concerned if you don't particularly understand the rules of, or strategy behind, chess – nothing I will say hinges on such an understanding.

To begin drawing out the features of formal systems, let's consider the chess board configurations depicted in Figure 7.1.

It is immediately apparent that the two boards are in different configurations, or *states*. Furthermore, we can all agree on a description of how the two depicted states differ: one of the white pieces has moved two squares towards the black pieces. We can be more precise than that though. If we label the horizontal from 'a' to 'h' left to right, and the vertical from '1' to '8' bottom to top, then we can say:

[1] The piece which was in square f2 in state A is in square f4 in state B.

We can, if we know about chess, add layers of interpretation to [1]. At the first level of interpretation we can say that a *pawn* which was in

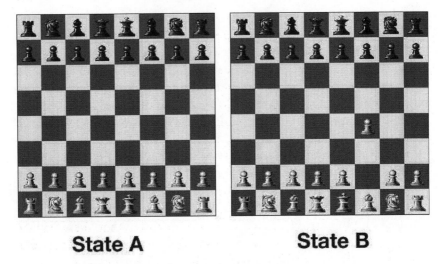

State A   **State B**

Figure 7.1   Two states.

square f2 in state A is in square f4 in state B. At the next level of interpretation we can say that the transition from state A to state B represents a *valid move* in chess – a move made according to the *rules* of the
game. We can also say that state A represents the beginning configuration, or *initial state*, of a chess game. At yet another level of interpretation we might say that the move depicted is an *interesting* or *dull* move.

However, none of this interpretation concerns us for the moment.
We just want to concentrate on [1] and use it to bring out some crucial
features of chess, without which the game could not be played – features so obvious that you have probably never had cause to reflect on
them.

What interests us about [1] is that given a chessboard in state A and
labelled as we have described, [1] carries all the information required
to recreate state B – even if we know *absolutely nothing* about chess.
In fact, we do not even need to recognise the states as configurations
of a chess board in order to apply the information in [1] to state A and
generate state B.

Let us recast [1] in terms of a task, or procedure, as:

[2]   Take the piece in square f2 and move it to square f4.

Presented with a labelled chess board configured in state A and told
[2], we could easily achieve the task without any understanding of
the task's *significance*, without any interpretation of its *meaning*.
Obviously, we need to interpret the meaning, in natural language, of
the words *describing* the task, but the task itself can be carried out

purely mechanically, without any appreciation of its significance. We will call such a task an *effective procedure*.

The application of [2] to state A to generate state B relies, for its *effectivity*, on the fact that states of a chess game are *effectively distinguishable*. The concept of effectivity will be doing some work in the next few chapters, so let's discuss those obvious features of chess I mentioned, by virtue of which states of chess are effectively distinguishable.

Firstly, there is no ambiguity about where pieces begin or end, nor about where squares begin or end. Pieces are discretely bounded spatial objects and squares have clearly delineated borders. Secondly, there is never any ambiguity about whether or not a piece is 'in' a square. These conditions are necessary in order that there can be *formal rules* which govern the legitimate ways in which pieces can move – a characteristic feature of chess and all such games.

To say that states of chess are *effectively distinguishable* is to say that there is an *effective procedure* which will decide the matter. You should be getting some sense by now of what an effective procedure might be – [2] is an example of one. Let's try to nail down a working definition. Let us call a procedure effective *iff* it can be achieved by merely following a specified set of steps – a list of instructions, a recipe – without any understanding of the significance or meaning of the task. This list of instructions we can call an *algorithm*. Another way of referring to the effectivity of a procedure is to say that it is *algorithmic* – that there is an algorithm for implementing the procedure.

The term 'algorithmic' has common-parlance usage – you may have heard someone call a task or chore purely algorithmic to describe that it is simply a matter of going through the motions, taking the prescribed steps – a mundane task. In other words, an effective procedure.

It is fairly clear that distinguishing states of chess is algorithmic. Let's specify the algorithm:

1. Begin with square a1.
2. Repeat step 3 sixty-four times unless instructed to halt. When instructed to move on to the next square, move to the adjacent square on the right-hand side if there is one, otherwise move to the leftmost square of the row immediately above.
3. If the square in state A is empty then:
   – if the corresponding square in state B is empty, move on to the next square, otherwise halt and utter 'the two states are different'.

If the square in state A is occupied then:
- if the corresponding square in state B is occupied by a piece of the same form, move on to the next square, otherwise halt and utter 'the two states are different'.
4. Utter 'the two states are formally equivalent'.

Although specifying this algorithm amounts to a rather tedious spelling-out of a task which is, for us, both simple and obvious to execute, it does drive home the point that I need bring no understanding to bear in order to achieve the task. I do not need to know that I am comparing states of chess. I do not need to know what chess is. I do not need even need to know what a game is. I need only follow the *formally specified* instructions.

You might think that this algorithm is not a good representation of how you conceive of yourself going about the task. When presented with state A and state B and asked to compare them, it is immediately obvious to us how they differ (without beginning comparison at square a1, moving on to a2, etc.), even though they differ only in respect of the positioning of one piece. This is partially because state A is a regular pattern on a sufficiently small scale. If the pieces were distributed more randomly on the board, no doubt we would find ourselves more closely following the algorithm above. If the board were ten times longer along each side – 6,400 squares in area – and there were ten times as many pieces, our only hope for a correct comparison of two states would be to follow the above algorithm.

In any case, the point here is merely that the task *can* be achieved by following an algorithm and, hence, is effective.

The final aspect of effectivity we need to appreciate is the *finitude rider* on effective procedures. A procedure only counts as effective if it can be carried out in *finite time*.

Counting the molecules in this book is an effective procedure. They are all presently configured in a solid and are pretty much sitting still, so we could, in principle, set up a very fine three-dimensional grid for reference and simply enumerate the molecules in each grid cube in order. It would take a very, very long time, but *finite* time.

Counting the natural numbers, on the other hand, is *not* an effective procedure. Although the process is mechanical – *name zero; name the successor of the last number you named; repeat the previous step until you can go no further* – it cannot be completed in finite time. Every natural number has a successor – there are infinitely many of them. So implementing this process is not effective.

Now that we have developed a good understanding of effectivity, it is time to put the concept to work in characterising formal systems.

7.2 STATES AND RULES

Formal systems are composed of two collections: a collection of *states* and a collection of *rules*. The specification of any given formal system consists in the specifications of its states and of its rules.

We can define states over any entities we choose, provided distinguishing between any two given states is effective.

To give a few examples, we can define states as configurations of a game board, as configurations of a finite array of switches, as distributions of people in a finite array of theatre seats, or as distributions of mail in a finite array of pigeon holes.

There must be only finitely many *entities* to define states over, otherwise any two states will not be effectively distinguishable. There can, however, be infinitely many *states* in the collection – many of the formal systems we will meet have infinitely many states, defined *recursively* over only a few entities. We will see how this trick is achieved in a moment.

All of the above examples of states employ physical objects as *tokens* – pieces, switches, people, mail – and states are distinguished by the discrete arrangements of these tokens. We could, however, just as easily use symbols on paper as tokens and define states over *strings* of these symbols (provided of course that the strings are effectively distinguishable). All of the formal systems we will be playing with in our survey of computational theory will be just such *symbol systems*.

Once we have defined a collection of states, we need to specify a collection of *rules*. Rules operate on states to *generate* other states. Rules, like states, are constrained only by considerations of effectivity. States can be defined over anything you like, provided any given two are effectively distinguishable. Similarly, rules can be anything you like provided they meet two constraints.

Firstly, determining whether a given rule applies to a given state must be effective. Typically not all rules will apply to all states. Secondly, if a rule applies to a state, it must effectively deliver a finite set of possible output states.

So rules take states, effectively modify them, and return distinct states.

Now that we know how to specify a formal system and what the constraints are on states and rules, let's exemplify with our first toy formal system.

7.3 SPECIFICATION

We want to discuss the properties and operations of formal systems, so let's define a *symbol system* to play with. Let's call it [STR] – it will be a *string system*.

States of this system will be finite strings of the symbols ■ and ◊. To give a few examples: ■■ is a state of [STR], ◊◊■■◊■◊◊■ is a state of [STR] and ◊ is a state of [STR].

It should be fairly clear that any two states will be effectively distinguishable. It may not be as clear that while there are only two *types* of symbols which we are defining states over, and while there can be only finitely many *tokens* of these in any given state, there are *infinitely many* states. This may be clearer in a moment when we give a *formal* specification of states using recursive definition; however, we first need to help ourselves to a few more concepts.

Firstly, we need the concept of an *initial state*. Specifications of formal systems often include an initial state, or beginning configuration of the system. All board games have an initial state – state A in Figure 7.1 represents the initial state of chess.

Secondly, there is a special and important state of this system (and typically of string systems in general) and this is the *empty string*. The empty string is precisely that – a string of no symbols at all. We will write it as ø. It is important to remember that the empty string *is a string*, and – by stipulation – is a state of [STR]. The need for the empty string will become apparent in a moment.

We will also need to employ *string variables* which we will write as Φ or Ψ. String variables stand for strings – any string you like (including the empty string). We need to make use of string variables and the empty string in order to formally specify rules of sufficient generality. Again, this will become apparent in a moment.

Finally, we need to represent *string concatenation*. String concatenation simply means joining two strings together – i.e. writing them one after the other. We will write the concatenation of two strings Φ and Ψ as $\Phi\Psi$ – i.e. their typographical concatenation. So, if Φ is standing for the string ■■◊ and Ψ is standing for the string ◊■ then $\Phi\Psi$ will be ■■◊◊■ and $\Psi\Phi$ will be ◊■■■◊.

Note that $ø\Psi\Phi = \Psi ø\Phi = \Psi\Phi ø = \Psi\Phi$. In other words, any string is identical to its concatenation with the empty string, regardless of where you put it. In simpler terms, tacking on nothing always leaves you with what you started with.

Before we give the complete formal specification of [STR], let's first just give an informal description of how the rules of this system will work.

[STR] will have only two rules. Rule one will say that we can take any state which begins with two boxes and ends with a diamond (and has whatever you like – including nothing – in between) and output a state which begins with a diamond and is followed by whatever came between the two boxes and the diamond in the input state (which may be nothing).

Rule two will say that we can take any state which has a diamond in it somewhere (and has anything you like – including nothing – before and after the diamond) and output a state which begins with two boxes, ends with a diamond, and in between has whatever it was that came before the diamond in the input state.

Explaining the rules informally like this is rather laborious. However, the formal specification is quite concise and tidy, as the following demonstrates.

[STR]

[S1]   ø is a state
[S2]   If Ψ is a state then so is Ψ■ and Ψ◊
[S3]   Initial state is: ■■◊◊

[R1]   ■■Ψ◊ →   ◊Ψ
[R2]   Ψ◊Φ   →   ■■Ψ◊

Now, let's carefully interpret this formal specification and make sure that it captures the system we have informally described.

The first thing to note about the formal specification is that, while giving an informal description of the system took many wordy paragraphs, the formal specification is given in five short lines. There is great economy of expression to be had in formalisations.

It may not be clear, on first examination, that [S1] and [S2] capture *all and only* the states of this system – this is an example of *recursive definition*. [S1] is the *base clause* – it simply stipulates that the empty string counts as a state. All the work is done in [S2] – the *recursive clause* – which says that if you take any state and tack on a ■ or a ◊ the result will be a state. A little thought should suffice to show that any finite string of the symbols ■ and ◊ can be constructed through repeated applications of [S2], given [S1]. [S3] merely stipulates the initial state.

The astute reader may wonder whether [S1] and [S2] capture *only* the states of the system – the worry being that they do not seem to explicitly rule out infinitely long strings. We need not be concerned, however – if you take any string of finite length and add one symbol, the result will always be a string of finite length.

We have already given informal readings of [R1] and [R2]. Work back and forth between those two paragraphs and their formal specification above until you are convinced that [R1] and [R2] do in fact capture the content of those paragraphs. The left-hand side (LHS) of the arrow of each rule describes the *form* of input states and the right-hand side (RHS) describes the *form* of output states.

It should be clear now why we want to use string variables. Using string variables lets us refer to a *class* of strings which share the same *form* (e.g. any string which begins with . . . and ends with . . .). String variables *always* refer to the same string in the LHS and the RHS of the rule – that is precisely the point in using them. Whatever you take Ψ to be in the input side of a rule must be the same in the output side of that rule. There are no restrictions on string variables *between* rules though: having taken Ψ to be one string in the application of one rule has no bearing on what Ψ can be in further applications of rules, or even in further applications of the same rule. If this is not yet clear it should become so in a moment when we examine the operations of the system.

It may still not be terribly clear why we want to be able to refer to the *empty string*, so let's demonstrate its use.

Rule [R1] will apply to the initial state ■■◊◊. In this case, what is in between the initial ■■ and the final ◊ is a single ◊, so in applying the rule we take Ψ to be ◊, in which case the output of [R1] will be ◊◊.

Another way of saying this is that the initial state is of the *form* required to apply [R1], namely the form ■■Ψ◊ (where in this case Ψ = ◊). Hence, if we apply the rule, we get an output state of the form ◊Ψ (where Ψ = ◊), namely ◊◊.

Rule [R1] will also apply to the state ■■◊ as it is of the form ■■Ψ◊ (where Ψ = ø). By similar reasoning [R2] will apply to the states ◊, ■◊ and ◊■. This is why we need the empty string.

Exercise 7.1

The paragraph above mentions four applications of rules to states: [R1] to the state ■■◊ and [R2] to the states ◊, ■◊ and ◊■.

What will the output of these applications be in each case? How are the string variables instantiated in each case?

Recall that typically, not all rules will apply to all states – determining whether or not a given rule applies to a given state must be effective. It should be obvious that neither [R1] nor [R2] will apply to every state of [STR]. It should be equally obvious that determining whether either rule applies to a given state is effective.

If you feel you have not quite followed all of the content of this section, then go back over the material until you are comfortable with it. We have met many new concepts in the last few pages and have started using symbolic representations. We will be building on this understanding in the pages to come so it is important for what follows that you first master the material to this point.

If you feel comfortable with the material we have covered and had no difficulty with Exercise 7.1 then it is time for us to go on and use the operations of [STR] to illustrate further concepts.

7.4 GENERATION AND DERIVATION

The operations of formal systems consist in successive applications of rules to states. Given an initial state, we can help ourselves to a distinction between states which will arise during the operations of the system, and those which, while they meet the *criteria* for possible states, never actually arise during the operations of the system.

Consider chess for example. States of the system are configurations of thirty-two (or fewer) tokens of twelve (or fewer) types – subject to certain restrictions – in an eight-by-eight array. There will, however, be a very large number of possible states which never arise during a game of chess. For instance, the state depicted in Figure 7.2 is impossible to achieve from the initial position given the rules of chess.

We will call a state a *generated state* if it can be obtained from the initial state through successive applications of rules. Generated states then will always be the output of some rule. Consequently, there is a simple effective procedure for *ruling out* a state as generated. If a state is generated, it will fit the output form of at least one rule, hence (by contraposition) if a given state does not fit the output form of any rule, it *cannot* be a generated state.

Unfortunately, determining whether a state *is* generated is not so straightforward. Fitting the output form of a rule does not guarantee that a state is generated – this is merely a *necessary condition* on generated states: its failure guarantees that we don't have a generated state, but its satisfaction does not guarantee that we do have a generated state.

If this is not obvious, here's an analogous situation. If I'm in Melbourne then I'm in Australia. So being in Australia is a *necessary*

Figure 7.2   A non-generated state.

condition for being in Melbourne. Hence, if I am not in Australia, I am guaranteed not to be in Melbourne. However, being in Australia does not guarantee that I *am* in Melbourne – I could be in some place outside Melbourne but within Australia, e.g. Brisbane.

When we are interested in formal systems, we are interested in determining whether or not certain states are *generated*. A generated state is one which can be *derived* in the system. To show that we can derive a state in a system is to give a *derivation*.

A derivation is a demonstration of the successive applications of rules to states, beginning with the initial state of the system and ending with the state we are interested in. Formally, a derivation is a finite sequence of lines, the first of which is the initial state of the system, the remainder of which are generated states, each obtained through the application of some rule to the state on the previous line.

Let's look at an example: suppose we wanted a derivation of the state ◊■■ in the system [STR]. Here's a derivation which does the job:

1. ■■◊◊   initial state
2. ■■■◊◊   [R2] Ψ = ■■◊   Φ = ø
3. ◊■■◊   [R1] Ψ = ■■◊
4. ■■◊■■■◊   [R2] Ψ = ◊■■ Φ = ø
5. ■■■■◊   [R2] Ψ = ■■   Φ = ■■◊
6. ◊■■   [R1] Ψ = ■■

The annotations on the right tell us how the state has been obtained – which rule was used and how the string variables in the rule have been *instantiated*. So, for example, state 2 was derived from state 1 (the initial state) by applying [R2] to it and taking Ψ to be ■■◊ and Φ to be ø (empty), resulting in an instantiation of the form ■■Ψ◊ (where Ψ is ■■◊), namely ■■■■◊◊.

But note that we could have applied [R2] to the initial state in a different way. We could have taken Ψ to be ■■ and Φ to be ◊, resulting in ■■■■◊. This means that [STR] is a *non-deterministic* formal system.

A system is *deterministic* if, for any given state, at most one rule applies to it and in only one way. If more than one rule applies to any particular state of the system, or if one rule applies to a particular state of the system in more than one way, then the system is *non-deterministic*.

Exercise 7.2

How many different ways can you apply [R2] to the state ■■■■◊◊ and what would the output be in each case? What about the states ■◊◊■◊◊ and ◊◊◊■■? What is the pattern?

You may have noticed that the result of applying [R2] to the initial state in the alternative way we discussed (taking Ψ to be ■■ rather than ■■◊) is identical to state 5 in the derivation. This means that we can actually get from the initial state to state 5 in the example in only one step (rather than four).

It will often be the case that there will be more than one way of deriving a given state. Often, we are interested in finding the simplest (i.e. shortest) derivation.

Exercise 7.3

Give the shortest derivation in [STR] of the state ◊■■.

You might wonder – given the indeterminacy of [STR] and the plurality of derivations – whether there is, in general, an effective procedure for finding derivations in a formal system. This turns out to be a very important question for the classical Artificial Intelligence research tradition and it will be our focus when beginning Chapter 11.

For the moment, a couple of informal heuristics will serve to guide you through the following exercises. Firstly, work backwards from the solution. Examine the state you want to derive and determine whether it could have been the output of any rule. If not, no need to continue; if so, you will have a guide as to which input(s) could have delivered that output. Try and get back to the initial state by working backwards through rules this way.

Secondly, aim for your goal. If you have a choice of rule applications to a state which is longer than your goal state and one choice results in a shorter output, it is likely (but not guaranteed) to be a good way to go.

Exercise 7.4

(a) Augment the system [STR] with the following rule:

[R3] ■Ψ◊ → ■ΨΨ◊

and give properly annotated derivations for the states:

1. ◊■■ 4. ◊◊◊■
2. ◊ 5. ◊■■◊■■■◊
3. ◊◊■◊ 6. ◊◊◊■◊◊

(b) Can you generate a state without a ◊? Explain your reasoning.

7.5 GENERATION TREES

The answer to Exercise 7.4(b) brings out an important feature of the system [STR] – that it has no *terminal states*.

We will call a state *terminal* if it is a *generated* state of the system to which no rules apply. So, while ■■ is a state of [STR] to which no rules apply, it is not a *terminal* state since it is not a generated state.

All generated states of [STR] are *ipso facto* the output of a rule and, hence, must contain a ◊. Since [R2] applies to any state which contains a ◊ in it anywhere, there are no generated states to which no rules apply. Hence, there are no terminal states of [STR].

Consider the specification below for the formal system [BIN]. Does this system have any terminal states?

[BIN]

[S1]   ø is a state
[S2]   If Ψ is a state then so is $\Psi 1$, $\Psi 0$ and $\Psi \omega$
[S3]   initial state is: ω

[R1] $\Psi \omega \Phi \rightarrow \Psi 1 \Phi / \Psi 0 \Phi / \Psi \omega \omega \Phi$

The single rule of this system has a choice of three possible outputs – the slash '/' represents 'or'. Read informally, it says that any ω in a state can be rewritten as a 1, or a 0, or as two ωs.

Given the initial state, the rule allows for the derivation of all and only the finite strings of the symbols 1, 0 and ω. In other words, there are no states of the system which are not generated. But are there any terminal states?

Consider the three ways in which we can apply the rule to the initial state. The resultant output states will be 1, 0 and $\omega\omega$ respectively. The first two of these do not contain the symbol ω. As the only rule in the system applies only to states which *do* contain at least one occurrence of the symbol ω, the states 1 and 0 are *terminal states*.

In fact, a little reflection serves to show that the terminal states of [BIN] will be all and only the finite strings of the symbols 1 and/or 0. In other words, all and only the finite strings of binary (e.g. 1001011) are terminal states of [BIN]. So any finite string of binary has a derivation in [BIN].

This is clearer if we draw up a *generation tree* as shown in Figure 7.3.

Turn the diagram upside down and it becomes a little clearer as to why these are called *tree* structures. The state at the top is the *root node*.

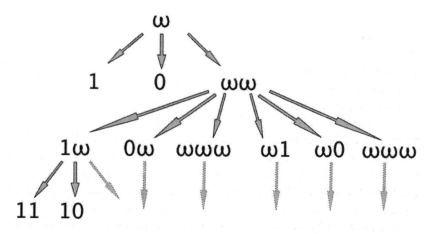

Figure 7.3   Generation tree.

An arrow leading from a *node* represents a possible application of a rule to the state at that node – the resultant output state of that rule application is at the node the arrow leads to. So the root node has three arrows leading from it as there are three possible ways to apply the rule to it. We will call these arrows *branches*, in keeping with the tree metaphor.

Some of the branches in Figure 7.3 lead to *terminal nodes*. A node counts as *terminal* if there are no arrows leading from it (i.e. if there is a terminal state at the node). Terminal nodes are the leaves, or *tips*, of our tree.

The first *iteration* of the system is represented by the first level of the generation tree after the root node. Figure 7.3 gives the complete generation tree for the first two *iterations* and a partial tree for the third *iteration*.

The first iteration has three nodes, two of which are terminal. In total, the first iteration contains two ωs – the two at the only non-terminal node. We know that there will be three ways of applying the rule to each ω so there will be six branches leading from the first iteration. In other words, there will be six nodes at the second iteration (as shown in the example).

Exercise 7.5

(a) The third iteration of the tree in Figure 7.3 is incomplete. Use the reasoning in the above paragraph to determine how many nodes there should be at the third iteration.
(b) How many of these nodes will be terminal?
(c) Find a large sheet of paper or a whiteboard and draw up a complete generation tree for [BIN] down to the fourth iteration.

Given a generation tree, we can read derivations straight off it by simply following a branch from the node we are interested in back up to the root node. For example, here is a derivation of the state '11' in [BIN] as read off the leftmost branch of the tree in Figure 7.3.

1.	ω	initial state	
2.	ωω	$\Psi = \emptyset$	$\Phi = \emptyset$
3.	1ω	$\Psi = \emptyset$	$\Phi = \omega$
4.	11	$\Psi = 1$	$\Phi = \emptyset$

So there is a general procedure for finding derivations in a formal system: simply complete the entire generation tree for the system, find the state you require a derivation for and read the derivation off the tree.

Unfortunately, as you would have seen if you attempted Exercise 7.5(c), generation trees can get very complicated very quickly. Many systems, including [BIN], suffer from *exponential explosion* – branches proliferate exponentially as we iterate.

What's worse is that, as you may have realised, some branches just go on for ever – they go infinite. This means that the procedure of drawing up the generation tree is not *effective* for systems whose trees have infinite branches. We will investigate the ramifications of this further when we discuss search procedures in Chapter 11.

Exercise 7.6

Give derivations in [BIN] for the states:

(a)   1001
(b)   0100101
(c)   000111

7.6 FORMALITY AND ISOMORPHISM

There is one last point to make concerning formal systems before we move on and do something more interesting with them. It is important to appreciate that the only important or relevant properties of formal systems are *formal* properties – properties of *form*.

For the purposes of the operations of a formal system, it is never important how the system is physically realised. Consider chess again. The pieces could be carved of wood or sculpted in stone, they could be symbols on paper or on an electronic screen, they could be coins or some other tokens pressed into impromptu service, they could even be people on a sufficiently large board.

The only features relevant to the distinguishing of states and the application of rules are the arrangements of the system in effectively distinguishable *forms*.

Another way of saying this is that the operations of a formal system are entirely independent of the medium (or substrate) in which they are *instantiated.*

This should remind you of the substrate independence claimed by functionalist theories of mind. For a functionalist, the only relevant things to know about mental states are *functions*. Similarly, when considering formal systems, the only relevant things to know about are *forms*.

The operations of a formal system are also entirely independent of any interpretation of the system. While formal systems are, in principle, interpretable (I can, for instance, interpret a whole range of instantiated formal systems as games of chess), I do not need to engage in any interpretive work in order to be able to apply rules to states – I need merely follow algorithmic procedures.

So, as is probably obvious to you by now, if am investigating some system [A] which has all and only the same formal properties of some system [B] then I *just am* investigating system [B]. If two systems are *formally equivalent* then they are instantiations of the *same system*. Whether I play chess with pieces, symbols, coins or people, I am playing chess.

If two systems are formally equivalent – if they have all and only the same formal properties – then we will say they are *isomorphic* to each other, or *isomorphisms* of the same formal system.

A formal system [A] is *isomorphic* to a formal system [B] iff we can derive [B] from [A] through uniform substitution of symbols. For instance, consider the system specified below:

[S1]   Ø is a state
[S2]   If X is a state then so is Xa and Xb
[S3]   Initial state is: *aabb*

[R1] $aaXb \rightarrow bX$
[R2] $XbY \rightarrow aaXb$

    where X *and* Y *are string variables*

It should be fairly clear that the above example is isomorphic to the original presentation of [STR]. In fact take any symbol you like and substitute it uniformly for *a*, and similarly for *b*, and the result will be another isomorphism of [STR]. The term 'symbol' can be interpreted quite broadly here to include physical tokens such as coins or people – we could, for instance, use ordered queues of men and women to investigate [STR] (provided we could effectively distinguish them).

The point of interest here is that for any formal system we might care to investigate, there will be an isomorphic symbol system. This is good news if we are interested in applying automated methods to the investigation of formal systems.

Now that we have a sufficient understanding of formal systems, their features and their operations, it is time to put formal systems to the use for which we have introduced them. In the following chapter, we will see how we can use a particular kind of formal system to do *computation*.

Let me say again that before you continue, it is important to have mastered the material to this point. We will continue building on this foundation in the next two chapters, so if there is anything of which you are uncertain, now is the time to revise. If, on the other hand, you are ready for some more challenging material, read on.

CHAPTER 8

COMPUTABILITY

The formal systems we have looked at so far have been very rudimentary string systems. Consequently, their useful application is rather limited. We can, however, employ formal systems to rather more interesting and useful ends. In particular, we can use formal systems to do *computation*. In this chapter, we are going to use a particular kind of deterministic formal system – a *register machine* – to rigorously define *computability*.

8.1 REGISTER MACHINES

Register machines are theoretical entities. They can, however, be physically implemented (as can any formal system). Modern digital computers as we know them are implementations of a special kind of register machine, as we will see in the following chapter.

Simple register machines, such as the ones we will examine in this chapter, can be straightforwardly implemented with piles of stones, coins or some other physical tokens. If you have a collection of coins or other tokens handy, it is highly likely to be useful to have them with you when working through the examples and exercises in this chapter.

The pigeon hole analogy is quite apt when first starting to think about register machines. Recall from section 7.2 that we can define states over any collection of entities we choose, provided any two given states are effectively distinguishable. In particular, we could define states as distributions of letters in pigeon holes. This is directly analogous to the way we want to define states of register machines.

States of register machines are contents of a finite sequence of *registers*. Registers are to be understood as discrete containers, hence the pigeon hole analogy. We will refer to these registers as R_0, R_1, R_2, . . . , etc. There may be an infinite number of registers; however, we make the simplifying assumption that only a finite number of them have contents at any given time.

The *content* of a register can be represented as a natural number – the number of items which it contains. So, if we have a pigeon hole with three letters, or a discrete pile of three coins, then we can say we have a register which contains *three* (items). Precisely what the items are (letters, coins, assorted objects, etc.) matters not at all. Any entities which can form effectively distinguishable states can serve as the symbols manipulated in a formal system.

A sequence of three pigeon holes containing, one, three and two letters respectively will be isomorphic to a sequence of piles containing one, three and two coins respectively. Both are isomorphic to a register machine with the contents, *one*, *three*, *two*, in the first three registers.

States of register machines, then, can be represented as finite sequences of natural numbers. The sequence 1, 4, 16, 2, 27 represents a register machine with 1 in R_0, 4 in R_1, 16 in R_2, 2 in R_3 and 27 in R_4. In other words, the numerical sequence represents an ordered sequence of five piles containing 1, 4, 16, 2 and 27 things respectively.

8.2 PROGRAMS

Register machines are formal systems. We now know what register machine *states* are – for all intents and purposes they are simply finite sequences of natural numbers. We next need to know what the *rules* are.

Register machines have only one rule. This rule takes a special form and is called a *program*.

A register machine *program* is a finite number of lines, each of which have two components: a *line number* and an *instruction*. Line numbers are simply natural numbers, assigned to facilitate reference to lines of the program. Each line must be assigned a unique line number. These are conventionally consecutive for readability, but need not be. *Instructions* take one of two forms: they are either *increment instructions* or *decrement instructions*.

An *increment instruction* is something of the form I *a b*, where the I stands for 'increment' and the *a* and *b* are numerical variables – they stand for natural numbers. A *decrement instruction* is something of the form D *a b c*, where the D stands for 'decrement' and, as with increment instructions, the lower case roman letters are numerical variables.

Recall that there must be effective procedures for both determining whether a rule applies to a state of a formal system and for applying rules to states of formal systems.

The effective procedure for determining whether a register machine program applies to a given register machine state is as follows:

1. Examine the contents of R_0 (this first register is always set aside for this purpose and is referred to as the *program counter* or *pc*).
2. If there is no line of the program which begins with the number in the pc then the program does not apply to the state.
3. If there is a line of the program which begins with the number in the pc then the program applies to the state.

The effective procedure for applying a register machine program is as follows:

1. If the line of the program which begins with the number in the pc contains an *increment instruction* (I *a b*), then *increment* (add one to) R_a and put *b* in the pc.
2. If the line of the program which begins with the number in the pc contains a *decrement instruction* (D *a b c*), then *decrement* (take one from) R_a and put *b* in the pc. If R_a is already empty (contains zero) then do nothing except put *c* in the pc.

The instruction forms I *a b* and D *a b c* are precisely that: *forms*. Register machine instructions are *instantiations* of these forms. Instantiated forms assign values to variables – in this case natural numbers to the numerical variables. So examples of actual lines of a register machine program look like this:

```
1  I  1  2
2  D  3  4  3
```

Line 1 instructs us to increment R_1 then put 2 in the pc (R_0). Line 2 instructs us to decrement R_3 *if we can* then put 4 in the pc, otherwise (if R_3 is empty) just put 3 in the pc.

So, when looking down the lines of a register machine program, the numbers in the first column after the instruction letters refer to registers – they tell us which register to increment or decrement. The numbers in the next column tell us which number to place in the pc after successfully executing the instruction. Numbers in the third column (which only appear in decrement instructions) tell us which number to place in the pc if we cannot execute the instruction.

Registers can always be incremented as there is no largest natural number. They cannot, however, always be decremented. The contents of registers are *natural numbers* not *integers*. The natural numbers include zero and the *positive* integers. If a register contains 1 it can be decremented once more then we say the register is empty. An empty register cannot be decremented. One can't take something away from a pile of nothing.

We can amalgamate the effective procedures for determining program applicability and applying a program into a single algorithm for *running* a program, as follows:

1. Look in the pc.
2. If there is no program line beginning with this number then halt.
3. If there is a program line beginning with this number, execute the instruction on that line.
4. Repeat.

So *running* a program simply involves repeated applications of first determining whether the program applies to the current state and then executing the relevant instruction until the program is no longer applicable (until there is a state with a number in the pc which has no corresponding program line).

8.3 RUNNING A PROGRAM

Now that we know what register machine states are and what it is to *run* a register machine program, let's examine the operations of a simple register machine. Consider the following program:

[ADD]

```
1  D  2  2  3
2  I  1  1
```

Let's exemplify the operations of this simple program and try to determine what it does. We know by looking at the first column after the instruction letters that there are only two registers referred to in this program – R_1 and R_2. So let's apply this program to an initial state which has contents in these two registers, let's say 3 in R_1 and 2 in R_2. Let us also suppose that our initial state contains 1 in R_0 (the pc) – this is conventional as it ensures that the first thing to happen will be the execution of line 1 of our program.

This initial state is represented on the first line of Figure 8.1. Each line below the initial state represents the resultant state of one application of the program.

When the program [ADD] is *run* with the initial state described, the sequence of operations will be as shown in Figure 8.1.

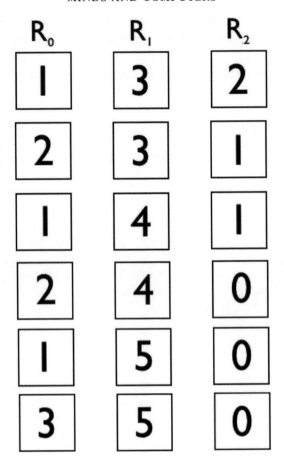

Figure 8.1   Sequence of operations.

If you have your pile of tokens handy, pull them out now and arrange three piles to represent the initial state in the example above. Now, let's work stepwise through the example. We will simply be applying the algorithm for running a program which I described at the end of section 8.2.

The first thing we do is look at the contents of R_0 (the pc) in the initial state. It contains 1 and there is a line of the program which begins with 1 so we execute the instruction on line 1. Line 1 contains a decrement instruction – it tells us to decrement R_2 (which leaves 1 in R_2) then put 2 in the pc. The resultant state of this first application is represented on the line below the initial state in Figure 8.1. To be explicit, we now have 2 in the pc, 3 in R_1 and 1 in R_2.

We now repeat the process. The pc contains 2 and we have a line 2 so we execute that instruction. The instruction on line 2 is an increment instruction which tells us to increment R_1 (which makes 4 in R_1) then put 1 in the pc. The resultant state is represented on the third line of Figure 8.1.

The pc now contains 1 so we execute the instruction on line 1 again, decrementing R_2 (which leaves 0 in R_2) and putting 2 in the pc. We then execute the instruction on line 2 again (since we now have 2 in the pc), incrementing R_1 (which makes 5 in R_1) and putting 1 in the pc.

Now we have 1 in the pc, 5 in R_1 and 0 in R_2. When we try to execute the instruction on line 1 again, we see we cannot. R_2 is empty so it cannot be decremented. Instead, we just put 3 in the pc.

When we then come to apply the program again, we see it is no longer applicable. We have a number in the pc with no corresponding program line so the program halts. The terminal state we reach is one in which R_1 contains 5 and in which R_2 is empty.

Exercise 8.1

(a) Run the program [ADD] a few times using different initial states. Each initial state should have 1 in the pc but choose whatever values you like for R_1 and R_2. Use pencil and paper if that is all you have but you may find it helps greatly to use piles of tokens.

(b) The program will always terminate with some value in R_1. Compare this, in each case, to the values which R_1 and R_2 took in the initial state. Can you ascertain what the program is doing?

8.4 COMPUTATION

Having completed Exercise 8.1 you would have noticed that the program [ADD] always terminates with a number in R_1 which is the sum of the numbers which were in R_1 and R_2 in the initial state. In other words, the program [ADD] adds the numbers in R_1 and R_2 and terminates with their sum in R_1.

The program does this by virtue of implementing a very simple algorithm for addition. If we have two piles of things to add, then, irrespective of how many are in each pile or which pile is the larger, we can find their sum by implementing the following effective procedure. First try to take one away from one pile. If you are successful, add one to the other pile. Repeat until the first pile is empty, at which point the sum of the two piles will be in the second pile. This is precisely what the program [ADD] does (where the 'first pile' is R_2 and the 'second pile' is R_1).

Given any two numbers x and y, the program [ADD] generates their sum. Another way of saying this is that the program [ADD] *computes* addition.

The previous paragraph marks the first appearance of the concept which it is the aim of this chapter to explicate, namely *computation*. We are on our way to a precise formal account of what it is to *compute*. Firstly, however, there are a couple of things we can say *informally* about computation.

For any given combination of a register machine program and an initial state, the sequence of applications of the program is a *computation*.

This is what register machines do – they *compute*. Precisely what it is they compute we shall come to in a moment.

We might note before we move on that there are some register machine programs which never terminate. Hence, there are some *computations* which never terminate. Consider the one-line register machine program which reads 1 I 1 1. This program will never terminate; it will simply continue to increment R_1 ad infinitum.

Unfortunately, there is no effective procedure for determining whether or not any given program will halt. This is the *halting problem* well known to computer programmers. There are certain pitfalls to be wary of in programming, particularly when dealing with nested loops, which will lead to a program never terminating. Programmers are taught to be attentive to this and there are methods and conventions for avoiding non-terminal loops. There is, however, no *algorithm* with which we can verify that any given program will terminate. Certainly there are software verification tools which, among other things, look for repeated patterns as evidence of *likely* non-termination. There are, however, ways to go infinite that do not involve repetition – think of the sequence of natural numbers, or the digits of π.

8.5 COMPUTABLE FUNCTIONS

We have so far introduced *computation* informally as the operations of a register machine program. We have also seen an example of something which is not *computable*. There is no effective procedure (a fortiori no register machine program) which will determine whether a given program will halt. Consequently something which cannot be computed is the question of whether a given computation will halt.

Precisely what things then *are* computable? Answering this question requires a formal characterisation of the notion of *computability* which in turn requires a tiny bit of mathematical terminology.

We need the formal notion of a *function*. This is a concept which everyone will be familiar with from grade school arithmetic.

A *function* $f(x_1, x_2, \ldots, x_n) = m$ is a mathematical correlation between some fixed number n of inputs and a unique output m.

Addition is a *function*. It takes a fixed number of inputs (two) and there is a unique output for any given input pair. Multiplication, subtraction, division, exponentiation and all the other basic arithmetical operations are also *functions*.

Some functions may be undefined for certain inputs. These functions are called *partial functions*, in contradistinction to *total functions* whose output is defined for all possible inputs (of the appropriate type).

Addition and multiplication are *total functions*. Their input types are real numbers and their output is defined for any possible pair of real numbers.

Division, on the other hand, is a *partial function*. It also takes real numbers as input but its output is not defined for certain input pairs (division by 0 is undefined).

The number n of inputs which a function takes is known as the *adicity* of the function. Addition and multiplication have an adicity of 2. The squaring function has an adicity of 1.

The noun *adicity* also has adjectival cognates with appropriate numeric prefixes. We would say, for instance, that squaring is a *monadic* function, and that addition and multiplication are *dyadic* functions. Less technically, we can also refer to these as *one-place* functions, *two-place* functions, etc.

We know have all the terminology we require to give a precise formal definition of what it is to *compute*, as follows.

If f is a function with adicity n, then program P is said to *compute f* if:

when P is run with 1 in the pc, $x_1, \ldots, x_n$ in $R_1, \ldots, R_n$ and all other registers empty then:

- If $f(x_1, \ldots, x_n)$ is undefined then the computation never terminates
- If $f(x_1, \ldots, x_n) = m$ then the computation terminates with m in R_1.

This is quite a dense definition so let's unpack it and consider some examples.

The first thing to note is that the objects of computation are *functions*. The definition above stipulates conditions under which we can say a given register machine program *computes* a function.

The definition is read as follows. Firstly, there is an initial condition which states that the program should be run with 1 in the pc, the n inputs of the function in the first n registers, and all other registers empty.

So, for example, to determine whether the program [ADD] computes the dyadic addition function $f(x_1, x_2) = x_1 + x_2$ we first satisfy this initial condition by clearing all the registers and putting 1 in the pc, x_1 in R_1 and x_2 in R_2. We then run the program.

Since addition is a total function, we know that its output will be defined. That is to say, there will be an m which is the result of computing the function. Hence, for the program [ADD] to be said to compute the addition function, it must, according to our definition above, terminate with $x_1 + x_2$ in R_1.

The program [ADD] does, in fact, always terminate with $x_1 + x_2$ in R_1 when the stipulated inistial conditions are met. Consequently, we can now say (with the authority of the demonstrated satisfaction of formal conditions) that the program [ADD] computes the dyadic addition function.

We now have the precise formal account of computation we wanted to develop. We can use this to define *computability* as follows.

A function is *computable* iff there is a register machine program which computes it.

So, to recap, the operations of register machines are computations. The objects of computation are functions – register machines compute functions. To *compute* a function is to satisfy the formal conditions laid out above. To say that a function is *computable* is to say that there is at least one register machine which will compute it.

8.6 BUILDING PROGRAMS

In the remainder of this chapter we will be developing methods for constructing register machine programs to implement algorithms. We are going to write programs to compute various functions. In each case, I will set an exercise and then work through a possible solution. Attempting the exercises before reading the solution will aid significantly in consolidating your understanding of this material.

Exercise 8.2

Write a register machine program which copies the contents of R_1 into both R_2 and R_3, leaving R_1 empty. Assume we begin with R_2 and R_3 both empty.

All the difficulty in writing register machine programs resides in determining the algorithm. Once we have determined the stepwise process which is guaranteed to deliver the result, translating this into register machine code is quite straightforward. This is why employing physical tokens is useful – they help us work stepwise through potential algorithmic procedures.

The algorithm for Exercise 8.2 is as follows. First try to take a stone away from the first pile. If you are successful, put one stone in the second pile and one stone in the third pile. When the first pile is empty, its contents will have been copied into both the second and third piles.

Translating this into register machine code gives us:

```
1   D   1   2   4
2   I   2   3
3   I   3   1
```

Exercise 8.3

Write a register machine program which copies the contents of R_1 into R_2 but preserves the contents of R_1. Assume we begin with R_2 and R_3 both empty.

Very often we will want to copy the value of a register into another register, but *without* losing the value from the first register. Since the only way to copy a value is to decrement the register containing it while incrementing the target register we always lose the contents of the first register. We can use the same process to put the value back into the first register again, but then we lose it from the target register and find ourselves back where we started.

The way around this dilemma is to use a *working register*. Very often, we will need to employ *working registers* in programs to store values during the computation.

In this case, what we need to do is copy the contents of R_1 into both R_2 and R_3, then copy the contents of R_3 back into R_1. R_3 is our working register – its only function is to keep a copy of the value initially in R_1 so that we can put it back once we have finished copying it into R_2.

We can simply extend our solution to Exercise 8.3 as follows:

```
1  D  1  2  4
2  I  2  3
3  I  3  1
4  D  3  5  6
5  I  1  4
```

The solutions to Exercises 8.2 and 8.3 do not, strictly speaking, compute 'copying' functions. Recall from our definition of what it is to compute a function that computations always terminate with the result in R_1. So the solution to Exercise 8.2 actually computes the function which maps any input onto the value $0 - f(x) = 0 -$ and the solution to Exercise 8.3 computes the function which maps any input onto itself $- f(x) = x$.

Our interest, however, is not in the functions which these programs compute but rather in the methods they employ. The method of copying without loss is one which will feature frequently in our programs. The point of the last two exercises has simply been to develop a useful piece of code which we can employ as a subroutine in further programs.

We will need to employ the method of copying without loss to solve the following exercise.

Exercise 8.4

Write a register machine program which computes the multiplication function: $f(x_1, x_2) = x_1 \times x_2$. Assume you begin with 1 in the pc, x_1 in R_1 and x_2 in R_2, but make no assumptions about the contents of other registers.

As always, the difficulty is in determining the algorithm. Multiplying two numbers x and y can be understood as accumulating x copies of y. So the algorithm we want to implement will try to decrement R_1

and, if successful, will add a copy of R_2 to a working register. When R_1 is empty, we want to move the accumulated value in the working register (x copies of y) into R_1 and terminate.

We know we need at least one working register to accumulate the result in. We are also going to need another working register to effect copying without loss from R_2 into the register accumulating the solution. That is to say, each time we are successful in decrementing R_1, we want to copy R_2 into a solution accumulator (say R_3) as well as into another working register (say R_4), then we want to put R_4 back into R_2 before attempting to decrement R_1 again.

We have been told not to make any assumptions concerning registers other than R_1 and R_2 which hold the inputs of the function. This means that the first thing our program should do is ensure that our working registers are empty. This is easily done with one program line per register.

So the algorithm we want to implement will do the following. First, ensure the working registers (R_3 and R_4) are empty. Attempt to decrement R_1. If unsuccessful, move R_3 (the solution accumulator) into R_1 and halt. If successful copy R_2 into R_3 and R_4, then move R_4 back into R_2 before attempting to decrement R_1 again.

Translating this into register machine code gives the following:

1	D	3	1	2
2	D	4	2	3
3	D	1	4	9
4	D	2	5	7
5	I	3	6	
6	I	4	4	
7	D	4	8	3
8	I	2	7	
9	D	3	10	12
10	I	1	9	

Exercise 8.5

Write a register machine program which computes the squaring function:
$f(x) = x^2$. Assume you begin with 1 in the pc and x in R_1 but make no assumptions concerning the contents of other registers.

Squaring is simply a special case of multiplication – the case where both multiplicands are identical. So x^2 is simply x copies of x.

One way to design a register machine program to compute the squaring function would be to make minor alterations to the algorithm for multiplication we employed in solving Exercise 8.4, as follows. As well as clearing R_3 and R_4, we also clear R_2. Then, before attempting to decrement R_1, we copy R_1 into R_2, using R_3 as a working register to copy without loss. We then follow the algorithm for multiplying R_1 and R_2.

There is, however, another algorithm we could implement. Consider the following register machine program:

```
 1   D   2   1    2
 2   D   3   2    3
 3   D   4   3    4
 4   I   2   5
 5   D   1   6    13
 6   D   2   7    9
 7   I   3   8
 8   I   4   6
 9   I   4   10
10   I   4   11
11   D   4   12   5
12   I   2   11
13   D   3   14   15
14   I   1   13
```

Exercise 8.6

The register machine program above computes the squaring function. What algorithm does it implement?

It turns out that to find the nth square number, we can simply accumulate the first n odd numbers. So 1^2 is 1, $2^2 = (1 + 3) = 4$, $3^2 = (1 + 3 + 5) = 9$, etc. Figure 8.2 sheds some light on why this is the case. The above register machine program for squaring implements an algorithm which takes the sum of the series of odd numbers up to the nth term (where n is the input of the function – the number we begin with in R_1).

Exercise 8.7

Design a register machine program which computes the function $f(x) = 2x^2 - 1$ (The solution to this one is up to you.)

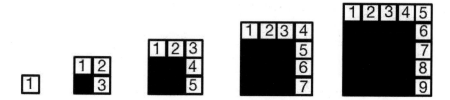

Figure 8.2   Progression of squares.

Before we move on, there are a few points of interest to draw from the material we have covered so far in this chapter.

We have now looked at register machine programs which can copy the contents of registers, compute addition and multiplication, and compute squaring in two different ways. The latter way of computing the squaring function involved generating, and accumulating, terms of a series.

All this has been built up from a very sparse set of basic resources – increment instructions and decrement instructions. We can see how these could be built up further into rather more complex computations – programs which implement highly complex algorithms. This is a notion we will revisit in Chapter 10.

Another point of interest is the demonstration that there can be more than one way of computing a function. In fact, very complicated functions are likely to be computable by a large number of algorithms. This is another notion we will revisit in Chapter 10.

In the following chapter, we are going to see how to use Gödel coding to facilitate reference to programs within programs. Using this method, we will see precisely how much can be achieved using only increment instructions and decrement instructions.

First, however, here is a final challenge exercise for those whose interest has been piqued.

Exercise 8.8 (Challenge)

We have seen how to accumulate the terms of a simple series. Rather than accumulating the terms, we could easily modify the program so as to return the nth term (where n is the input of the function), in other words to terminate with the nth odd number in R_1 rather than the sum of the first n odd numbers.

The Fibonacci sequence progresses as follows:

0, 1, 1, 2, 3, 5, 8, 13, 21, 34, 55, 89, . . .

Let fib(x) be the function whose output is the xth Fibonacci number.

Write a register machine program to compute fib(x).

Hint: fib(1) = 0;   fib(2) = 1;   fib(x + 1) = fib($x-1$) + fib(x)

CHAPTER 9

UNIVERSAL MACHINES

This chapter completes our exposition of computational theory. So far, we have discussed formal systems broadly and we have used a special kind of formal system – a *register machine* – to define *computation* and *computability*.

In doing so, we gained some insight into what is computable and what is not. We saw, for instance, in section 8.4 that there is no effective procedure for determining whether or not a given program will halt. Since register machines are bound by constraints of effectivity, we can say that there is, therefore, no register machine which can determine, of any given program, whether or not it will halt. Consequently, since computability is defined in terms of register machines, we can say that the *halting problem* is not computable.

We also know from section 8.5 that the objects of computation are *functions* and we have a precise formal definition of what it is to *compute* a function. So we know that at least *some* functions are computable and that *only* functions are computable.

There are still, however, important questions left unanswered. For one thing, we have no account of the *limits* of computation. We do not yet know precisely *which* functions are computable.

We are also still left wanting of a formal account of *effectivity* and *algorithmicity*. Recall from section 7.2 that we defined effectivity only *informally* in terms of algorithmicity. We said that a procedure is effective, just in case there is an algorithm for carrying it out. However, the notion of an algorithm was also only fleshed out informally, in terms of a mechanical procedure or a set of instructions which can be carried out without any understanding or interpretation.

The first thing we will do in this chapter is to give a precise answer to the question of the limits of computation by tying our *informal* notion of effectivity to our *formal* account of computability, thereby delivering a formal definition of algorithmicity.

With this account in hand, we are then going to develop a description of a single machine which can compute *any* computable function – a *computer*.

9.1 CHURCH/TURING THESIS

The *halting problem* is a particular instance of the *Entscheidungsproblem* – or *decision problem* – which was of interest to mathematicians and logicians well before there was a formal theory of computation. The decision problem for a particular formal system refers to the question of whether or not there is an effective procedure for determining, of any given state of the system, whether or not it is *generated* in the system. If there is such a procedure, the system is said to be *decidable*.

For the sake of only those readers who may have an understanding of modern logic, let me say the following. The question of whether or not a particular logical system is *decidable* is the question of whether or not there is a method for showing, for any arbitrary sentence of the language of the system, whether or not that sentence is provable in the system. However, the concept of *provability* reduces to the concept of *generation*. If we take the initial state of the system to be its axioms and the proof theory of the system to be its rules, then decidability simply amounts to whether or not one can show, of any arbitrary sentence, whether or not it is included in some state derived from the initial state in accordance with the rules. Those readers who have done a first course in logic will know that the propositional calculus is decidable, but the predicate calculus is not.

Do not be concerned if the previous paragraph means little or nothing to you. It is merely intended as an aside for those with some exposure to logic.

It was in the course of arguing that a particular logical system is not decidable (the predicate calculus for those for whom that means something) in his 1936 paper – *On Computable Numbers, with an Application to the Entscheidungsproblem* – that the British mathematician Alan Turing formally defined *computability* and, in doing so, laid the foundation for what has become computer science.

Turing defined computability in terms of theoretical machines which are now known as *Turing machines*. It is far more common in the computational theory literature to see talk of Turing machines than it is to see talk of register machines. Turing machine computability and register machine computability are, however, provably equivalent – they can compute all and only the same functions.

Wherever you see talk of Turing machines, or Turing computability, you can substitute this with register machines and register machine computability.

I chose to employ register machines for an initial presentation of computational theory for two reasons. Firstly, simple register machines are readily amenable to physical implementation with a collection of tokens (coins, buttons, etc.). This can serve as a great aid to understanding their operations. Secondly, while register machines and Turing machines are functionally equivalent, the mechanics of register machine operations are closer to the way computation is actually implemented in computers as we know them

In his 1936 paper, Turing argued that all and only those procedures which were algorithmic could be computed by Turing (register) machines.

In another paper published in 1936, an American mathematician, Alonzo Church, argued that the informal notion of effective calculability could be understood (at least for calculations over positive integers) in terms of the formal notion of a *recursive function*. In both cases, Church and Turing aimed to tie a notion which was well understood informally – that of algorithmicity, or effectivity – to a concept for which there was a precise formal definition.

The Church/Turing thesis, as it is now commonly referred to, can be expressed quite concisely as: *all and only effective procedures are computable functions.*

It is important to appreciate that the use of 'computable' in the Church/Turing thesis refers to a particular formal understanding of computability, namely *register machine computability* as we have defined it. This may not seem terribly important but there is considerable scope for misinterpreting the Church/Turing thesis if one does not keep this in mind and it is common to see such misunderstanding in the literature.

It is also worth pointing out that the equivalence between effective procedures and computable functions is a *thesis* and not a *theorem*. In other words, it is a proposed equivalence but it is not a *proven* one. There seems, however, by the very nature of the thesis, to be little chance of actually proving it, since it equates a formal notion with an informal one. One direction of the thesis – that all computable functions are effective procedures – is obvious since we have defined computability in terms of effective procedures. The other direction – that only computable functions are effective procedures, or that all effective procedures are computable functions – is less obvious. It is the case, though, that since 1936 we have amassed considerable evidence in favour of the thesis and run across no counter-examples to it.

So the Church/Turing thesis provides us with answers to both of the questions we identified at the beginning of this chapter. We now know precisely which functions are computable – all and only those which are algorithmic. We also now have a *formal* characterisation of effectivity/algorithmicity – precisely that which we gave for computability in section 8.5.

According to the Church/Turing thesis, there is a register machine program for any given algorithm (since all effective procedures are register machine computable). Add to this the fact that register machine programs are deterministic formal systems, and that the operations of any given deterministic formal system are algorithmic, and we can see that *algorithmic procedures, effective procedures, register machine programs, deterministic formal systems* and *computable functions* are all equivalent ways of speaking – they all specify the same class of things.

9.2   GÖDEL CODING

We have seen that for any algorithm one can specify, we can design a register machine program to compute it. The final aim for this chapter is to specify a single register machine program which can *itself* compute any algorithm. In other words, we want to specify a register machine program which can *emulate* or *instance* any other register machine program.

This means we are going to need some way of making register machine programs suitable to be the input of another register machine. Given that the inputs of register machines are natural numbers, we are going to need some way to *code* register machine programs as natural numbers. Furthermore, this coding procedure needs to be unique and effective. In other words, as well as being algorithmic, the coding procedure must be such that any given register machine program results in a unique code, and vice versa. The procedure we are going to use is known as *Gödel coding*.

Kurt Gödel was a brilliant logician and mathematician who delivered some of the most important results in mathematical logic in the twentieth century. Gödel's incompleteness theorems, delivered in a 1931 paper, are arguably the theorems which are the most well known by those who are not mathematicians or logicians, and simultaneously the least understood. There is, in fact, rather a lot of nonsense written about the Gödel results in philosophy of mind.

Unfortunately, a proper understanding of the Gödel incompleteness theorems, such that we might see how they are frequently misinterpreted in arguments against computationalism, would require considerably more mathematical sophistication than one would expect from the introductory reader to whom this volume is directed. Suffice it to say, if you run across arguments against computationalism employing these theorems, they should be taken with a grain of salt.

The method of Gödel coding is one of the very neat tricks Gödel used to prove the Incompleteness Theorems. It facilitates reference to elements of a formal system from within that system, which is precisely what we require. Fortunately for our purposes here, understanding the method of Gödel coding requires no more mathematical sophistication than an understanding of the operations of multiplication and exponentiation, and of the notion of a prime number.

The mathematical operations of squaring and cubing are examples of *exponentiation*, which is the more general operation of raising one number to the *power* or *exponent* of another. The expression a^b is evaluated by multiplying a with itself b times – a is the *base* and b is the *exponent*.

A *prime number* is a number which has no factors other than itself and 1. An interesting feature of primes – and a crucial one for our purposes – is that every natural number can be uniquely expressed as a product of primes. The proof of this is a little complicated so you'll have to simply take my word for it here. Furthermore, prime factorisation is algorithmic (there are a number of known algorithms). Again the details are complicated so you'll just have to take my word here too.

So for any given natural number, we can effectively and uniquely determine its prime factors. The method of Gödel coding exploits this by coding programs as exponents of primes.

Recall that a register machine program is a sequence of instructions of one of two forms – either the form I a b or the form D a b c.

For increment instructions we need to code three pieces of information – that it is an increment instruction and the numbers a and b. For decrement instructions we need to code four pieces of information – that it is a decrement instruction and the numbers a, b and c.

So to code increment instructions, we use three prime numbers. We begin with 2 which will indicate that the code is of an increment instruction. We then multiply this by 5 raised to the power of a and 7 raised to the power of b.

To code decrement instructions, we use four prime numbers. We begin with 3 which will indicate that the code is of a decrement instruction. We then multiply this by 5 raised to the power of a, 7 raised to the power of b and 11 raised to the power of c.

To express this symbolically:

I a b is coded as $2 . 5^a . 7^b$ (where the . represents multiplication) and

D a b c is coded as $3 . 5^a . 7^b . 11^c$

(We will always leave these expressions in exponential form.)

Note that the code of an increment instruction will always be an even number, the code of a decrement instruction will always be divisible by 3, and any number which is not divisible by 2 or 3 is not the code of any instruction.

If Φ stands for some instruction, then we will use #Φ to refer to the code of Φ. So, to give a few examples:

if $\Phi = $ I 2 3   then #$\Phi = 2 . 5^2 . 7^3$
if $\Phi = $ I 7 4   then #$\Phi = 2 . 5^7 . 7^4$
if $\Phi = $ D 3 8 5 then #$\Phi = 3 . 5^3 . 7^8 . 11^5$
if $\Phi = $ D 2 6 3 then #$\Phi = 3 . 5^2 . 7^6 . 11^3$

and so on.

Exercise 9.1

Code the instruction bodies of the solutions to the exercises in chapter eight. (Disregard the line numbers for the moment – just code the instructions.)

Now that we know how to code instructions, we are going to use exactly the same procedure to code entire programs.

A program is some number n of lines consisting of a line number and an instruction. To simplify things, we make the assumption that lines are numbered in the sequence of natural numbers beginning from 1. When we decode program codes into their constituent instruction codes, we simply enumerate them in the same sequence. This allows us to disregard line numbers in codes.

Given a sequence of n instructions, we can code this by simply multiplying the first n primes, each of which is raised to the exponent which is the code of the relevant instruction. So the first prime is raised to the exponent which is the code of the first instruction, the second prime is raised to the exponent which is the code of the second instruction, and so on.

To express this symbolically:

given a program 1 Φ_1, 2 Φ_2, 3 Φ_3, . . . , n Φ_n
it can be coded as $2^{\#\Phi 1} . 3^{\#\Phi 2} . 5^{\#\Phi 3} . . . p^{\#\Phi n}$ (where p is the nth prime).

This might seem rather complicated but some examples will serve to show that it is actually very straightforward. We always simply leave the expressions in their exponential form and we assign variables to stand for instruction codes to make things easier to read, as the examples below illustrate.

The program:

1 D 3 1 2
2 D 1 3 5
3 D 2 4 5
4 I 3 2

is coded as:

$2^a . 3^b . 5^c . 7^d$

where:

$a = 3 . 5^3 . 7^1 . 11^2$
$b = 3 . 5^1 . 7^3 . 11^5$
$c = 3 . 5^2 . 7^4 . 11^5$
$d = 2 . 5^3 . 7^2$

The program:

 1 D 3 1 2
 2 D 4 2 3
 3 D 1 4 9
 4 D 2 5 7
 5 I 3 6
 6 I 4 4
 7 D 4 8 3
 8 I 2 7
 9 D 3 10 12
10 I 1 9

is coded as:

$2^a . 3^b . 5^c . 7^d . 11^e . 13^f 17^g . 19^h . 23^i . 29^j$

where:

$a = 3 . 5^3 . 7^1 . 11^2$
$b = 3 . 5^4 . 7^2 . 11^3$
$c = 3 . 5^1 . 7^4 . 11^9$

$d = 3 \cdot 5^2 \cdot 7^5 \cdot 11^7$
$e = 2 \cdot 5^3 \cdot 7^6$
$f = 2 \cdot 5^4 \cdot 7^4$
$g = 3 \cdot 5^4 \cdot 7^8 \cdot 11^3$
$h = 2 \cdot 5^2 \cdot 7^7$
$i = 3 \cdot 5^3 \cdot 7^{10} \cdot 11^{12}$
$j = 2 \cdot 5^1 \cdot 7^9$

Exercise 9.2

The second example above codes the solution to Exercise 8.4. Code up the solutions to the other exercises in Chapter 8.

Although we are leaving program codes in exponential form, they resolve to natural numbers. Admittedly their resolutions are very large natural numbers, but they are natural numbers nonetheless. As such, we now have a unique and effective procedure for coding any register machine program of any length as a single natural number which is uniquely and effectively decodable. This means we now have a mechanism for referring to programs within programs, since a register can hold the code of a program.

Exercise 9.3

Give three examples of numbers which are, for different reasons, not program codes.

Exercise 9.4 (Challenge)

What is the smallest natural number which is a program code?

9.3 A UNIVERSAL MACHINE

Armed with the method of Gödel coding, we are now ready to describe the register machine program which can compute any algorithmic function.

Let [UM] be the register machine whose program is described by the following procedure.

First, decode the contents of R_1. If R_1 does not contain the code of a program then halt. If R_1 does contain the code of some program P, then add 1 to all the registers referred to in P, then run P and on termination copy the contents of R_2 to R_1.

It is clear that this procedure is effective, hence, by the Church/Turing thesis, there *is* such a register machine program.

If we were to run [UM] with the input #P, $a_1, \ldots, a_n$ in the first n + 1 registers, it would deliver the same result as running a machine with program P and the input $a_1, \ldots, a_n$ in the first n registers.

This is why the procedure for [UM] involves adding one to all the registers in P (if R_1 decodes to #P) and why we copy R_2 to R_1 on termination of P – because the inputs of P will all be displaced along one register in [UM].

Consequently [UM] can emulate any register machine program. We simply put the code of that program in the first register and the n inputs of the function in the subsequent n registers and run [UM] as described.

The register machine [UM] is known as a *universal machine* – it can compute any register machine computable function. More simply, [UM] is a *computer*.

Modern digital computers, as we know them, are instantiated universal machines. Personal computers, mainframes and even super-computers are no more powerful than [UM] – there is nothing that they can compute that [UM] cannot. In fact, [UM] is *more* powerful than any physical machine since it is a theoretical idealisation which is unconstrained by physical manifestation. We will have more to say about this in the following chapter.

We have now completed our survey of computational theory. Equipped with our new understanding of what computers are, it is time to return to philosophical material and discuss the theory of mind which our central concern is to evaluate – computationalism.

CHAPTER 10

COMPUTATIONALISM

So far, we have considered the question of what minds might be and in doing so, have examined a number of philosophical theories which aim to provide an answer to this question. We suspended that discussion after introducing functionalism and turned our efforts to developing a rigorous account of what computation is. It is now time to pick up where we left off.

In this chapter, we are going to use the understanding of computational theory we built up in the previous section to flesh out the functionalist framework in a particular way.

There are various ways in which one can be a functionalist, depending on how one analyses the functions of mental states. Our aim is to give a fair and precise treatment of the theory that these functions are to be fleshed out in *computational* terms.

In doing so, the utility of having developed an account of computational theory will become apparent. In the first instance, we can now speak of computation without simply engaging in loose talk – we have a precise formal definition and some subtle distinctions at our disposal. This allows us to see that certain objections sometimes raised against computationalism do not actually target the theory – they target *straw men* by virtue of an insufficiently sophisticated understanding of what *computers* are.

To target a *straw man* – or to commit a *straw man fallacy* – is to characterise an opposing position as being weaker than it actually is and to then argue against the weaker position. Arguing against a weaker misconception does no work at all against the actual opposing position. Hence, if one commits this fallacy, one is said to build a straw man simply for the purposes of knocking it down.

We begin with a clear characterisation of computationalism. Following this, our first priority is to address, and clear up, some possible misconceptions of the theory. We are then going to discuss

94

precisely what a computationalist is committed to and consider some immediate implications.

In this chapter, we will also discuss some of the merits of computationalism and deal with some prima facie objections to the theory. Evaluating computationalism more fully will be the concern of the remainder of the book. We are going to see how disparate material from the cognitive disciplines bears importantly on the tenability of the theory.

Let us now address the question of precisely what is involved in claiming that mental states are computational states.

10.1 WHAT COMPUTATIONALISM ISN'T

Computationalism is the view that to have a mind is to instantiate a particular formal system or collection of systems. In other words, since mental operations are held to be the operations of formal systems, mental operations are held to be *computations*. So to have a mind, claims the computationalist, *just is* to be engaged in certain computational processes.

Computationalism is clearly a species of functionalism. The functionalist holds that states are mental solely by virtue of their characteristic functions in mediating relations between inputs, outputs and other mental states. Computationalism is simply a way of fleshing out these mediating relations – the relations in question are held to be computations.

Computationalism is *not* the view that the operations of formal systems per se are mental operations. That is, it is not the view that instantiating any formal system at all is sufficient for having a mind. This clearly overcommits the computationalist as it would require them to attribute mentality to all manner of artefacts – thermostats, traffic lights, handheld electronic games – in a patently ludicrous fashion.

So in fairness to the computationalists, let's be clear that they are committed only to the view that instantiating a *particular* formal system – let's call it [MIND] – is sufficient for having a mind.

This is rather a strong formulation of computationalism. A computationalist might hold that there is no single overarching formal system to be identified but, rather, that mentality is a function of some number of distinct algorithms. I want to advance a particular understanding of mentality – which I take to be fairly intuitive – which assumes a single overarching formal system, so I shall continue to work with this strong thesis until further notice. Do be aware though that the version of the

theory I am describing is not the only story available to a computationalist. If, however, a strong version of the theory turns out to be
defensible then, a fortiori, any weaker version is defensible.

The computationalist, then, is not committed to the view that the
operations of your personal computer are mental operations. Nor is
she committed to the view that very powerful computational devices,
such as supercomputers, have minds. She is committed only to
holding that should some substrate run the program [MIND] then
that substrate thereby has a mind.

I have referred to [MIND] in three different ways now – as a formal
system, as an algorithm and as a register machine program. Recall
from Chapter 9 that by the Church/Turing thesis, these three are
equivalent ways of speaking. If [MIND] is an algorithm, it can be
implemented by a register machine program (which just is a deterministic formal system).

Let's clear up another possible misconception of the theory. The
computationalist claim differs in another important way from the
view that personal computers can or do have minds. Modern digital
computers, as we saw in the last chapter, are instantiated *universal
machines*. The computationalist is *not* claiming that to have a mind is
to instantiate a universal machine. She is claiming that to have a mind
is to instantiate a particular *register* machine – namely [MIND].

We need to be careful on a couple of points here. Firstly, we need
to appreciate that, while digital computers, as we know them, are
instantiated universal machines, they are *imperfectly* instantiated.
Universal machines are theoretical devices whose resources, while
finite by stipulation, are otherwise unlimited. Instantiated universal
machines are physical devices which are bound by physical constraints. So while universal machines can *in principle* run any program
(a fortiori can run [MIND]), instantiated universal machines are
limited *in practice* by their physical constraints and, as such, may not
have sufficient computational resources at the hardware level to run
certain programs, such as [MIND].

So there is a sense in which it is not quite correct to say that modern
digital computers are instantiated universal machines, as there may
be programs beyond their computational resources. Digital computers, as we know them, are *approximations* to universal machines.
Successive generations of computational hardware provide closer and
closer approximations as they provide greater and greater computational resources.

Even perfectly instantiating a universal machine in a substrate
would not in itself be *sufficient* for that substrate to have a mind. The

substrate must then run the right program, since having a mind, says the computationalist, is having the program [MIND] *in operation.*

We can give a fair measure of *practical computational power* along two parameters – storage space and processing speed. A given program's requirements can be said to exceed the practical computational power of a given device if its requirements exceed either the storage capacity or the speed of computation of the device (or both).

So while the computationalist is committed to holding that any universal machine (without constraints) can run [MIND] – and would thereby have a mind – she is not *ipso facto* committed to holding that any existent digital computer could have a mind. A computationalist might hold that the requirements of [MIND] exceed the practical computational power of (some or all) currently available (non-biological) computational devices.

Consequently a computationalist need not hold that your personal computer could have a mind. In all likelihood they will hold that it could not by virtue of its physical limitations. A computationalist will hold to the view that the human brain provides the biological computational hardware for implementing [MIND] in humans. Given what we know of the extraordinary storage capacity and speed of operation of the human brain, the computationalist is likely to argue that any non-biological computational device powerful enough to run [MIND] must have (at least approximately) the storage capacity and speed of operation of human brains. Digital computers, as powerful as they are becoming, are still not even close.

There is one final point of possible confusion to clear up before we move on. Recall from section 7.1 that for a procedure to be effective, it must, in principle, be able to be carried out, given sufficient time, by a human using only piles of stones (or paper and pencil) and bringing no understanding to the task. The operations of universal machines are entirely effective, so one of the things a human mind can do is approximate a universal machine (albeit with considerable constraints) by actively working stepwise through the operations of any given register machine program. Consequently, one of the things that a device running [MIND] must be able to do is approximate a universal machine (in at least the same fairly weak fashion in which humans can).

This does not, however, mean that approximating a universal machine is *sufficient* for having a mind. Quite the opposite. It means that irrespective of the status of computationalism, having a mind is sufficient for (very weakly) approximating a universal machine.

Let's recap the points of possible confusion we have covered so far. Firstly, a computationalist is not committed to the view that any computation is a mental operation. They are committed to the view that *particular* computations – those which are the operations of [MIND] – are mental operations.

Secondly, a computationalist is not committed to the view that instantiating a universal machine is sufficient for having a mind. They are committed to the view that a perfectly instantiated (unconstrained) universal machine has the *capacity* to have a mind. Since one of the things a mind can do is approximate a universal machine, the computationalist is also committed to the ability of any computational device running [MIND] to weakly approximate a universal machine.

Finally, a computationalist is not committed to the view that any given computational device could instantiate [MIND], as the program may have requirements which exceed the practical computational resources of the given device. Consequently, a computationalist can happily deny that a device such as your personal computer – an approximation to a universal machine – could ever have a mind. The computationalist is committed, however, to holding that any physical device with sufficient practical computational power to run [MIND] does have the capacity to have a mind. Precisely what computational resources are required by [MIND] is a matter for empirical discovery.

Now that we have carefully identified several possible misconceptions of computationalism, we can see that certain arguments against the theory which trade on these misconceptions are unsound.

For instance, the following argument should clearly not be licensed:

P1   Computationalism says that all mental operations are computations.

P2   My personal computer performs computations.

∴   Computationalism says that the operations of my personal computer are mental operations.

P3   But my personal computer clearly does not have a mind.

∴   Computationalism is false.

The premises P1 – P3 are not in dispute. P2 and P3 are clearly true and computationalism does make the claim attributed to it in P1. The argument goes wrong in the transition from P1 and P2 to the interim conclusion. The inference is not truth-preserving; the interim conclusion is, in fact, false.

We can prove that the inference is not truth-preserving by giving counter-examples to the form of inference employed, since truth-preservation is a matter of logical form (more on this in Chapter 15). The inference is of the logical form: C *claims that everything which is* A *is* B; *x is* B; *therefore* C *claims x is* A. Instantiate A as 'in Melbourne' and B as 'in Australia' (and anything you like for C and *x*) and we have a clear counter-example to the validity of this argument form – a demonstration that the truth of the premises does not guarantee the truth of the conclusion.

We can clearly see though how a misinterpretation of P1 could lead us to infer the interim conclusion, given P2. Computationalism does claim that all mental operations are computations but the converse, as we have seen, does not hold. Consequently, the fact that something performs computations does not guarantee that it performs mental operations. Were we to mistakenly read P1 as its converse (that *only* mental operations are computations – which is to say that *all* computations are mental operations), we would be led, erroneously, to believe that the above argument instances a valid form.

Exercise 10.1

Construct unsound arguments against computationalism which trade on the other possible misconceptions we have considered. In each case, explain why the argument is unsound.

If there is a false premise (attributing a claim to the computationalist) explain why the premise is false in light of a correct understanding of computationalism.

If there is an invalid inference in the argument, give a counter-example to the validity of the logical form employed and explain how a particular misconception would lead one to believe that the argument instances a valid form.

10.2 SOFTWARE AND WETWARE

Computationalism is often described as a 'software' view of the mind. The human brain is seen as providing the biological computational hardware – or *wetware* – which confers on humans the capacity to have a mind. Having a mind, on this view, is a matter of having the right program running in one's wetware.

This provides us with a solid methodological framework for investigating mentality. What we should be interested in, if computationalism is correct, is determining the program(s) for various kinds of mental processes with a view to building up [MIND].

Computationalism has been widely embraced in the cognitive disciplines and plays a large role in informing research programmes. Each of the empirical cognitive disciplines approaches the overarching goal of investigating mentality in distinct fashion, commensurate with their disciplinary methods and assumptions. In each case, however, a commitment to computationalism confers a broad methodology for pursuing these questions. Researchers who endorse the computational hypothesis will aim to deliver accounts which are in principle computationally implementable. That is to say, they will aim to develop accounts of mental processes as effective procedures. Very often, this will involve collaboration with computer scientists in developing computational models of mental phenomena.

This methodology has sparked off the research tradition known as *artificial intelligence*. There are weaker and stronger interpretations of 'artificial intelligence'. The *weak artificial intelligence* research programme simply involves aiming to construct artefacts capable of instantiating particular functions which are held to be (albeit weakly) constitutive of intelligence. This is the kind of 'artificial intelligence' which is often used to sell white goods.

The *strong artificial intelligence* research programme is of significantly more interest and of central concern in this volume. It commits to, and pursues, the possibility of developing artefacts which *have minds* in the sense that we take ourselves to have minds.

There are certain mental capacities which appear to be unique to human minds. These include the ability to reason complexly and abstractly about such things as mathematics, logic and metaphysics, and the ability to use language. Both the rational capacity and the linguistic capacity implicate a number of what we might call lower-order cognitive processes, such as the abilities to discriminate, to learn and to remember. These lower-order processes are achieved to a greater or lesser extent by other animals. The higher-order cognitive functions of abstract reasoning, language production and language comprehension, however, are uniquely human and, as such, will serve for our purposes as prime determinants of the kind of intelligence we attribute to humans.

Consequently, in the following chapters, we are going to concentrate on various methods of attempting to develop computational devices with rational and linguistic capacities.

We will also have much more to say about the conditions under which we might *attribute* mentality to an artefact in the final section of this chapter when we discuss the Turing test. Before we get to that, however, there is more to be said here about computationalism. To begin with, we can draw out some advantages of computationalism from our discussion so far.

Firstly, and most obviously, computationalism fleshes out the functionalist framework. Recall from Chapter 6 that we found the functionalist account somewhat wanting as a 'black box' view of mentality. Computationalism tells us what is going on inside the black box, namely *computation*. Consequently, computationalism confers a clear methodology for investigating mentality – we should be aiming to provide computational accounts of cognitive capacities.

Secondly, computationalism allows us to specify the relation between the mind and the brain by employing a useful wetware/software distinction. On this view, minds are what brains do. In other words, brains provide the computational resources to run [MIND].

We might also note at this point, that computationalism retains the substrate independence which functionalist theories enjoy and, consequently, is compatible with a purely material view of the mind without falling prey to either the multiple realisability objections which frustrate Australian materialism or the methodological vacuousness of token physicalism.

Computationalism, then, appears to enjoy the strengths of other theories of mind without being subject to the worst of their weaknesses. There are, however, a number of objections which we might mount against the view that all mental operations are computations. In the following sections, I will consider a number of prima facie objections – the kind typically raised against computationalism on first presentation. In each case, I will demonstrate how a computationalist might defend the theory against the objection in question. We will leave consideration of more sophisticated philosophical arguments against computationalism until Chapters 17 and 18.

10.3 VARIATION

The objection from variation runs as follows. Computationalism says that humans have minds by virtue of implementing [MIND], but human minds vary greatly. How can this be, given all minds are held to be isomorphisms of the same formal system?

It is certainly the case that there is considerable variation among individual minds. We all have different beliefs, desires, emotional responses and mental capacities. A computationalist will, of course, admit this – to deny it would be foolishness. This does not answer the objection, however. To do so requires further reflection on the notion of isomorphism.

Formal systems involve variables which are assigned values in any particular instantiation of the system. For two systems to be isomorphic to each other, they must be *formally* equivalent. This does not mean, however, that every isomorphism of a formal system must assign all and only the same values to variables. In fact, this will very rarely be the case.

When doing Exercise 8.1, you were instantiating various isomorphisms of the formal system [ADD] in order to ascertain the function which the program computes. In each case, you assigned different values to the initial input variables.

Further, if we take two isomorphisms of [ADD], both of which were assigned identical initial values, and then compare the contents of registers at different stages in their operations, we will see different values being held in each register. The operations of each system, however, are still formally equivalent – they are still isomorphisms of [ADD].

Now consider the system [MIND]. This must be an extraordinarily complex system with very, very many variables. Presumably, every instantiation of [MIND] is bound to be in a distinct stage of operation. Presumably also, every instantiation of [MIND] will begin with (albeit perhaps slightly) distinct assignments of values to variables.

Hence it should come as no surprise that any two given instantiations of [MIND] will vary greatly in terms of the values which are currently assigned to variables of the system.

It seems, at least prima facie, that beliefs and desires play the same functional role in all minds. That is not to say that my belief that (whatever) performs exactly the same functional role in my mental life as does your belief of the same content in your mental life. Rather, it is to say that beliefs, *qua* beliefs, have a functional role in deliberation, planning, motivation to action and so on.

Because beliefs and desires interact in a highly complex fashion in these mental functions, we should not expect that my belief that (whatever) will result in the same action as your belief of the same content. We should expect, however, that, *qua* belief, both of our beliefs function in deliberation, planning, etc.

Now, a computationalist is likely to argue that the *content* of beliefs, desires and the like – that which they are *about* – is to be understood in terms of the assigning of values to variables. In the system [ADD], the *content* of R_2 (whatever it might be) always functions in the same way – it is the value which is decremented while we increment R_1. Similarly, says the computationalist, the *content* of a 'belief-box' in [MIND] will always function in the same way, even though its content (what the belief is about) might be markedly different in different instantiations of the system or at different stages in the operation of the same system.

Consequently, a computationalist can hold that minds are isomorphisms (functional equivalents) of [MIND], yet still happily concede that all minds will vary, in small measure or large, in terms of the contents of beliefs, desires and the like. They can also readily explain variation among isomorphisms of [MIND] in terms of the particular actions which are partially determined by beliefs, desires, etc. with particular content, as the causal determinants of action will be held to be complexly interrelated with the values of many variables (the contents of many beliefs, desires, etc.). We will have a lot more to say about the idea of mental *content* in Chapter 18.

This line of argument does not speak directly to the fact that minds differ also in terms of capacities. For instance, some minds are more amenable to formal reasoning, some minds are more skilled at employing language, some minds are skilled in working with engineered artefacts, and some minds are capable of producing extraordinary works of art. This leads us to our next objection.

10.4 LEARNING

Minds can learn. Consequently, different minds can do different things. In other words, some minds can perform functions which other minds cannot (or can perform certain functions better than most other minds). How can computationalism account for this?

We have seen how computationalism can account for variation between minds with respect to their content; however, minds also vary with respect to capacities. This presents the computationalist with two further challenges. Firstly, she must give a computational account of the acquisition of new capacities, i.e. specify algorithms which govern learning.

Ideally, we would like a specification of the algorithm(s) (or varieties of algorithms) which *actually do* govern learning in human minds. However, for the purposes of defending computationalism against this objection, it suffices to show that learning is in principle effective, in which case the specification of any correctly functioning algorithm(s) will suffice.

Secondly, the computationalist needs to explain how there can be functional equivalence between two systems with different capacities, i.e. how it is that two isomorphisms of [MIND] can *do* different things.

· The latter challenge is the more fundamental so I will respond to this first. Let's reflect on what it is for two formal systems to be isomorphic to each other. In section 7.6 we said that a formal system [A] is isomorphic to a formal system [B] iff we can derive [B] from [A] through uniform substitution of symbols. So if, for instance, we take a chess board in some configuration and replace all the pawns with identical coins, then the result is another isomorphism of the same formal system.

Further, if we have two chess boards and we play one for a number of moves, then we will have two isomorphisms of the same formal system in different states. A move might then be available on one board – e.g. castling – which is not available on the other board.

The claim that two formal systems [A] and [B] are isomorphic amounts to the following: for all states x and rules R, if the application of R to ([A] in state x) generates state y, then the application of R to ([B] in state x) will generate state y.

Understanding that isomorphic formal systems in different states may yield different outputs given the same sequence of rule applications allows for a straightforward explanation of variation among minds with respect to their response to a particular situation. If two minds have different beliefs, desires, etc. then their [MIND]s are in different states, hence we should not expect the same rule application in both [MIND]s would result in the same output.

We have also gone some way towards understanding variation among minds with respect to capacities – different rules might be applicable to isomorphisms of [MIND] in different states. There is more to be said here, however.

Consider the system [UM] from Chapter 9. Recall that [UM] operates by decoding the value in R_1 and running the program that the value codes (if it codes a program at all). Now let's consider a similar register machine – we'll call it [OS].

[OS] will have a large number of registers set aside in which to store its own values. It will also have a large number of registers available in which to do computations on these values. Some of the values stored by [OS] will be codes of algorithms (programs). These can be executed, using other values of [OS] as program input, in the space set aside for computations and their output can then be stored as further values of [OS].

So, for instance, one of the registers of [OS] might contain #[ADD] (the code of the program [ADD]). At some point in its operation, the program [OS] might address this register and instantiate [ADD] using the values of two other registers as input, and store the output in another register.

Now, let us suppose that the program [OS] governs the operations of many algorithms (stored as values in its registers), that many of these may be in operation at any given time, that the output of algorithms can be employed as the input of other algorithms, and that the output of algorithms can determine which algorithm will be executed next (and with which values).

The program [OS], while it may sound more complicated than [UM], is clearly effective. When we speak of many algorithms being in operation *at the same time*, what we mean is that many algorithms are in process; however, only one step of one algorithm will be carried out at each time step. [OS] is therefore a register machine program like any other and, hence, is computable by [UM].

The system [MIND] can be understood as a very complex version of [OS]. It functions by governing the operations of a large number of algorithms which individually perform mental functions – algorithms which transform sensory data into perceptual representations, algorithms which govern bodily movements, algorithms which govern linguistic production and comprehension, algorithms which determine actions based on beliefs, desires and the like, and so on.

To employ the software analogy, [MIND] is best understood as a kind of *operating system* which manages the highly interrelated operations of a large number of applications and controls the hardware in which it is instantiated.

Some of the algorithms which [MIND] employs serve as *learning algorithms*. We will call an algorithm a *learning algorithm* if, as a result of employing it, the system is conferred with greater capacities.

There are at least two ways in which a mind can be conferred with greater capacities – it can gain new information or it can attain, or improve, skills or abilities. In the former case, the mind is learning *that* things are the case. In the latter, the mind is learning *how* things are done.

In the system [MIND] learning *that* things are the case corresponds with storing new content – i.e. storing new values in registers (where these values code content). Learning *how* things are done corresponds with storing new algorithms or with optimising existing algorithms.

Now, given our understanding of [MIND] as a kind of [OS] which is in continual iterated operation and which is expandable – i.e. has algorithms which store content and which generate or optimise further algorithms, both of which confer on the system greater capacities – we can help ourselves to an explanation available to the computationalist of variation among [MIND]s with respect to capacities.

Given that, as already established, any two [MIND]s are highly likely to be in distinct states, and given that, depending on the amount of time it has been operating for and the nature of the inputs it has received, a [MIND] will contain more or less stored content and greater or fewer available stored algorithms (optimised better or worse), it should come as no surprise that any two [MIND]s will vary significantly with respect to capacities.

This explanation responds to one of the two challenges presented to the computationalist by the learning objection – it explains how two isomorphic formal systems can have different capacities and, hence, how minds can be held to be isomorphisms of [MIND] despite immense variation among minds with respect to what they can do.

The challenge remains, however, for the computationalist to specify algorithms which govern learning. This is a challenge to which we will return at various points in the following chapters, particularly in Chapter 13 when we discuss automated reasoning systems, and in Chapter 19 when we examine learning in artificial neural networks. We will also be investigating the way in which humans learn languages in Chapter 16 and considering evidence that this learning is rule governed.

We have now considered two objections one might mount against computationalism. These were essentially stronger and weaker versions of the same objection – minds vary. In both cases, we have seen how a computationalist might reasonably respond. Let's consider one further objection against the theory that minds are computational devices.

10.5 CREATIVITY

Another standard prima facie objection appeals to the human creative capacity, as follows. The operations of formal systems are entirely mechanical but minds are *creative*. Minds create great works of art, music, architecture and literature, and have an enormous capacity to innovate. This characteristic creativity of human minds seems to be compelling evidence against computationalism which seeks to account for mentality in terms of purely mechanical operations.

It is certainly the case that it seems that nothing could be further from an algorithmic process than painting an artwork or composing an orchestral symphony. As we saw in Chapter 2 however, the way things seem is no reliable indicator of the way things are.

The challenge here for the computationalist is to explain how the mental functions we cite as paradigmatically 'creative' can be algorithmically delivered, contra-intuition. For that explanation, we need an understanding of this notion of creativity.

The opponent of computationalism might endorse a definition of creativity along the lines of: *an activity is creative if its result is the production of a work (an artwork, composition, etc.) which could not have been produced by simply following a rule-governed procedure.*

Although this definition is somewhat intuitive – after all we're all fairly certain da Vinci wasn't painting by numbers when he painted the *Mona Lisa* – it begs the question rather straightforwardly against the computationalist. Whether or not creativity can be accounted for algorithmically is precisely what is at issue.

So what is it about creativity such that our initial intuition is to contrast creative behaviour with rule-governed behaviour?

Well, firstly, not everyone is equally creative. People have different capacities for engaging in creative enterprises. We have already seen, though, how a computationalist can account for variation with respect to capacities so this is not sufficient as an objection, but it points us in the right direction.

It seems that what disposes us initially against the notion of creative behaviour being rule-governed is an intuition that were it rule-governed, it would be more readily teachable. It is characteristic of those we laud as creative masters – artists, artisans, composers, etc. – that there is something mysterious to others about their talent. Further, it seems that when it comes to creative endeavour, one either 'has it' or not. Certainly one can learn various techniques and methods for working with materials to generate certain effects; however, it is not clear how one could learn to 'be creative' per se.

This does not, though, speak against the possibility that this kind of behaviour is indeed underwritten by computational processes. Certainly the opponent of computationalism may well demand an account of how such behaviour could be computationally delivered; however, there are certain responses available.

For instance, the computationalist might tell a story about certain algorithms requiring certain computational resources for their implementation, such that variation with respect to the ability to acquire certain algorithms is to be explained in terms of variation in the substrate (brains) in which [MIND]s are realised.

In any case, the computationalist is well within their rights to dissent from answering the question – *how is creative behaviour computationally delivered?* – until this notion of 'creative behaviour' is more rigorously specified. In the absence of an independently plausible account of creativity which speaks directly against the possibility of such behaviour being rule-governed, the computationalist is no worse off in this respect than other theorists.

We have now considered a number of preliminary objections against computationalism and have seen how, in each case, the objection fails. We have yet to consider more sophisticated arguments against the theory. We shall hold off on these until after we have seen some artificial intelligence applications which will provide context for philosophical objections.

To recapitulate, we have discussed some common misunderstandings of computationalism and some bad arguments against the theory which trade on these misunderstandings. We have advanced a particular version of computationalism according to which the mind is analogous to a computer operating system, governing the operations of applications and controlling the hardware in which it is instantiated (the central nervous system). We have also witnessed the failure of several first-blush objections to computationalism.

The final thing to do in this chapter before we move on to examine various artificial intelligence applications is to spend some time reflecting on the conditions under which we are prepared to attribute mentality.

10.6 ATTRIBUTING MENTALITY

How do we determine whether or not something has a mental life? When we are not in the grip of philosophical scepticism, most of us are quite convinced that those around us have mental lives – at the very least we habitually act as if they do. Given that we have privileged access to only our own mental states, what is it about other people that leads us to believe that they have minds?

Well, for starters other people are physically very similar to ourselves and appear to be the same *kind* of being. As such, we expect them to share various properties with us, such as being capable of similar locomotion and interaction with the environment. This extends to an expectation that other people also experience an inner mental life. Clearly, however, such physical similarity is not *sufficient* for an attribution of mentality such as we experience, as these conditions might be met by non-human animals and very young children who, we judge, lack the kind of rich mental life we enjoy.

Nor indeed is such similarity with respect to physical capacities *necessary* for attributions of mentality, as we can imagine someone who clearly has a mind, in the strong sense in which we hold ourselves to have minds, yet lacks these physical capacities.

So what conditions do other people meet such that we attribute mentality to them? The key here – unsurprisingly given the professed focus of the coming chapters – lies in our use of language.

The competent and sophisticated use of language is a hallmark of the mental. By 'competent' and 'sophisticated' here I mean that the speaker is able to comprehend and produce novel utterances, use various linguistic devices to achieve various effects, and discourse on various topics involving various degrees of abstraction.

It is by virtue of evidencing this kind of linguistic capacity – which I might point out implicates a rational capacity – that we are prepared to attribute mentality to those things which *behave* in this way. Even if this capacity is realised in highly non-standard ways we seem to be prepared to take this as *sufficient* for an attribution of mentality. No one is going to deny, for instance, that Stephen Hawking has a mind, and a sterling one at that.

We need to be careful here – I have made a sufficiency claim but not a necessity claim. There are certain cases of aphasia (language deficit) such that sufferers will fail any standard test of linguistic capacity yet are still quite capable in other respects to the extent that we would certainly attribute to them a robust mentality. We will revisit this point when we discuss human language in Chapter 16.

This sufficiency of the linguistic capacity for mentality provides us with a rough and ready test for the presence of a mind. If the test subject can satisfy certain conditions we stipulate, which are designed to probe the subject's capacity to use language, then we might think that should be sufficient for assuming, at least as a working hypothesis, that the subject indeed has a mind.

In real life, this assessment is an ongoing enterprise. We tend to assume by default that others have minds and we revise our estimation of their mental capacities in accordance with their evidenced linguistic capacity. A little reflection on our social interactions serves to demonstrate this so I shan't argue any further for it. Certainly it is the case that it is through language that we are able to investigate the minds of others (here I intend 'investigate' in lay terms). The simplest way to find out about the beliefs, desires and so forth of others is to talk to them (although this is of course fallible).

It is precisely this intuition that Alan Turing seized upon in his seminal article *Computing Machinery and Intelligence*, wherein

Turing determines to investigate the question of the conditions under which we would be prepared to attribute mentality to an artefact.

Turing posited a test, which is standardly known as the *Turing test*, the successful passing of which he proposed as sufficient for holding the subject to have a mind.

The *Turing test* is often glossed in a weaker form than Turing intended, so let's employ some care in setting it up. The test is an extension of an old parlour game known as the *imitation game*. In the imitation game, a man and a woman are placed in separate rooms with an interrogator in a third. The interrogator employs an interme-diary and, by passing notes (which the intermediary reads out to the interrogator), interrogates the man and the woman for a specified length of time. The interrogator is allowed to question the man and the woman on any topic she sees fit. The point of the game is for the interrogator to try and determine which room contains the man and which the woman, and for the man and the woman to attempt to fool the interrogator.

Turing proposed that an adaptation of this game could serve as a barometer of mentality for artefacts (computers). In all fairness to computing devices, we should not expect them to satisfy conditions of physical similarity, or to be able to perform certain physical tasks in order to be said to have a mind. After all we have seen that this is neither necessary nor sufficient. We should expect, however, that if an artefact can satisfy us of its linguistic competence, then this would be a very good indication that the artefact has a mind.

The Turing test adapts the imitation game – in ways that are prob-ably already obvious – to provide a fair means of assessing the linguis-tic capacity of an artefact. It is conducted as follows. We put a computer in one room, a human in another, and a human interrogator in a third. The interrogator is able to communicate with the computer and with the human via a keyboard and monitor. The interrogator has an allotted amount of time to question each participant on any topic she sees fit, during which she attempts to discern which of her inter-locutors is the human, while both the human and the computer attempt to convince her that *they* are the human. If, at the end of the allotted time, the interrogator is unable to discern the machine from the human, we should say of the machine that it has a mind.

Exercise 10.2

What are your initial reactions to the Turing test as an indicator of machine intelligence? What problems do you envisage with using this method for determining the presence of a mind?

Keep in mind that the claim is *not* that passing the Turing test is *sufficient for having* a mind. The thought is that passing the Turing test gives us *good grounds* to suppose that the test subject has a mental life.

As it stands the Turing test remains underspecified. We have not yet set a length of time for conducting the test and a lot hangs on this. If, for instance, the test is to be conducted for only five minutes, then we might rightly have qualms about its reliability as an indicator of mentality. If, on the other hand, the interrogator is allowed to spend as much time as she requires until she feels capable of making an adequate assessment, then it seems we can have a lot more confidence in that assessment – particularly if, as we should, we take the assessment to be revisable, contingent on the results of further such tests.

Since 1991, an annual competition for the Loebner Prize (named after the philanthropist sponsor) has been conducted, in which participants submit to a formalised version of the Turing test. The competition is held over one day with a $100,000 prize and a gold medal on offer for the a machine which is held by the judges to be indistinguishable from a human being. As yet no machine has come close to this. Each year, however, a $2,000 prize and a bronze medal is awarded to the machine which performs the most impressively of the entrants.

Turing envisaged, in 1950, that by the year 2000 machines would be able to convincingly pass a Turing test of moderate duration. No doubt he would be disappointed with the current state of play – the transcripts of conversations with Loebner Prize entrants show that some rather unsubtle questioning suffices to distinguish the machine participants in very short order.

This is not, however, *ipso facto* an indictment of the Turing test, but rather of the current state of play in artificial intelligence.

Perhaps a behavioural test of this kind, while a good indicator under certain strong conditions, simply does not resolve the question satisfactorily one way or another in problem cases. In such instances it is not obvious what other methods we might employ to determine the presence or absence of a mind.

Exercise 10.3

Would a version of the Turing test involving a number of judges and unrestricted in duration serve as a reliable indicator of mentality? Can you think of any other conditions or constraints we might build in to the test to increase its reliability?

Let's pause now and think about where we are and where we are going. So far in this book, we have introduced various philosophical problems associated with understanding the mind and examined a number of theories which propose answers to these questions. We have also introduced some rudimentary neuroanatomy to get a feel for how brains work, given that there is clearly some important relation between brains and minds.

Of the philosophical theories of mind we have covered, functionalism was the least problematic and most satisfactory. However, it left us wanting to know more about the relevant functions in question. In order that we might be well prepared to understand a particular way of answering this, we then rigorously developed a precise account of computation.

Armed with this understanding, we have now examined a particular way of fleshing out functionalism which supposes that the functions in question are computations – namely computationalism. We have also lent some consideration to the question of the appropriate conditions under which we should attribute mentality. There will be much more to say about this as we progress.

In the coming chapters we are going to see how a commitment to computationalism confers a methodology for artificial intelligence research. We will be concentrating on various computational problems which must be solved in order to equip a device with the capacity to reason and use language.

Along the way we will be gathering evidence which bears on the tenability of computationalism – an issue we will explicitly return to at the beginning of Chapter 17 when we begin to mount serious challenges and sophisticated philosophical objections to the theory.

Now it is time to begin telling the story of artificial intelligence – a story which begins with the concept of *search*.

CHAPTER 11

SEARCH

Given the computationalist hypothesis as we have described it, investigating mentality involves investigating the operations of the formal system [MIND] and its constituents.

The classical or *symbol systems* approach to artificial intelligence involves trying to determine the algorithms for the cognitive functions involved in [MIND] and investigating their associated formal systems.

Many of the problems in the classical artificial intelligence tradition reduce to determining whether or not a certain state can be generated in a particular formal system, or to finding a particular generated state of a system. In other words, many artificial intelligence problems reduce to the problem of *searching* for a particular state.

In this chapter, we are going to briefly run through various methods for searching the generation tree of a formal system. You may wish to refer to section 7.5 to refresh your understanding of generation trees and the associated terminology.

11.1 TOP DOWN, BOTTOM UP

One method of determining whether a state of a formal system is generated, or finding a derivation for a particular state of the system, is to construct the entire generation tree for the system. If the state we are interested appears on the tree it is a generated state and we can read off its derivation(s) by following the branches back up to the root node.

For instance, suppose we are investigating the system [BIN] from section 7.5 and we are interested in whether the state 01 is generated. We can construct a generation tree for [BIN] beginning from the initial state and working our way down through all the possible ways in which rules can be applied to each node. Figure 11.1 gives a generation tree for [BIN] which shows that (and how) the state 01 is generated.

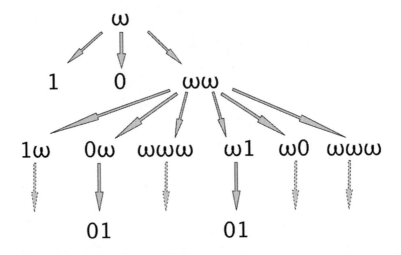

Figure 11.1   Top-down search.

A search of this kind is known as a *top-down* search. Often, however, it will be more efficient to begin with the state we are trying to derive and work backwards through the rules to see if we can get back to the initial state. This kind of search is called a *bottom up* search. Figure 11.2 depicts a bottom up search in [BIN] for the state 01.

One advantage of using a top-down search is that it is comprehensive. Every generated state appears on a completed top-down tree. Bottom-up searches, however, have the advantage of delivering all and only the possible derivations for the solution state, if there are any.

A major consideration in determining whether or not a top-down search or a bottom-up search will be more effective is the associated *branching factor*. The branching factor of a tree – the average number of descendants below each node – determines the complexity of the tree so ideally we want to minimise it. As Figures 11.1 and 11.2 demonstrate, the branching factor of a top-down search in a system will often be distinct from the branching factor of a bottom-up search. In the case of [BIN], a bottom-up search proves to be more efficient.

Exercise 11.1

What determines whether the branching factor of a bottom-up tree for a system will be lower or higher than the branching factor of a top-down tree?

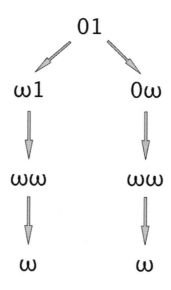

Figure 11.2   Bottom-up search.

11.2 BREADTH VERSUS DEPTH

Having decided between a top-down search and a bottom-up search, we then have to further decide on a procedure for searching the tree. Keep in mind that any formal system of sufficient interest for artificial intelligence researchers to be investigating will be considerably more complex than the toy examples we are looking at here. Consequently, choosing an appropriate search procedure can have a significant impact on the computational resources required to carry out the search. In fact, as we shall see, it can mean the difference between a successful search and a search which never halts.

A *breadth first* search involves exhaustively searching all the nodes at a given level before descending to the next level. Figure 11.3 shows the order in which we would search the nodes of a tree if we were conducting a breadth first, left first search.

A clear advantage of breadth first search is that it is exhaustive. If there are solution states to be found anywhere on the tree, the search will find them. A downside of breadth first search, however, is that it can be much more computationally expensive than necessary. This is particularly the case if the solution state is a long way down one of the branches.

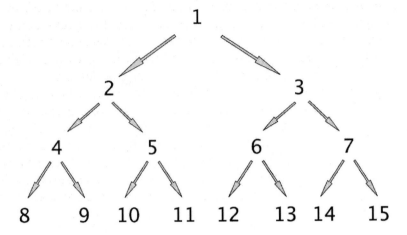

Figure 11.3 Breadth first search.

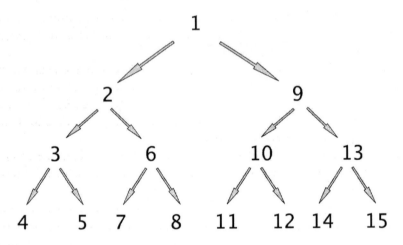

Figure 11.4 Depth first search.

If we expect there to be numerous solution states a long way down the branches, it is more efficient to conduct a *depth first* search. A depth first search involves searching all the way down the length of a branch until we either find a solution state or hit a terminal state. If we reach a terminal node, we backtrack only until we can descend an unsearched branch.

Figure 11.4 shows the order in which we would search the same tree from Figure 11.3 if we were conducting a depth first, left first search.

While depth first search can be notably more efficient in situations where there are numerous solution states deep down the branches of a tree, there is a danger in using it. If any of the branches of the tree go infinite – as they often will in interesting formal systems – and we use a depth first search, we run the risk of our search never terminating. If there are no solution states on a branch which goes infinite, our depth first search will continue down the branch ad infinitum.

A useful strategy can be to *combine* depth first and breadth first search. If we know, for instance, that some branches of the tree we are interested in will go infinite and we suspect that there will be numerous solution states distributed across nodes a long way down the tree, we might do a depth first search *only to a certain level*. If we exhaust the nodes down to that level without finding a solution state (which is tantamount to having conducted a breadth first search to that level) then we continue our depth first search to a deeper level.

11.3 HEURISTIC SEARCH

Breadth first and depth first are both what we call *blind* searches – they are conducted without any consideration of the *closeness to solution* of the nodes being searched.

Frequently, however, it will be the case that we have some effective procedure for determining, of a given node, the likelihood of there being solution states among the descendants of that node. A function which applies such an algorithm to nodes and assigns a value to them accordingly is a *heuristic function*.

Determining good heuristics is very often the most difficult element involved in solving complex problems in the symbolic artificial intelligence tradition.

Consider, for instance, an exemplar classical artificial intelligence project: determining an algorithm for playing chess well. One heuristic for playing chess would be to simply construct the generation tree for chess, then for any given game state – represented by a node of the tree – assign a value in accordance with the number of descendant nodes which represent winning states. Playing the game is then just a matter of always moving to the position whose node has the best value of those available.

Given a rule establishing an upper limit on the number of allowable moves in a game (such as the fifty move rule), we can keep the tree finite, so this procedure seems feasible until we begin to quantify the sheer complexity of chess.

The generation tree for chess suffers from drastic combinatorial explosion. There are four hundred nodes at the second level alone. By the time we get to the tenth iteration – by which time each player has made only five moves – there are around nine billion nodes. At the twentieth iteration there are something in the order of 10^{30} nodes and that is a *very* big number.

Even if we could generate one hundred trillion states per second, it would still take around ten billion years to produce the generation tree for only the first ten turns of chess.

Clearly then, the naive heuristic we described is computationally untenable. Similarly, blind search techniques – while they have their uses – are often computationally untenable for formal systems of sufficient complexity to be of interest to artificial intelligence researchers. Investigating the generation trees of exponentially complex formal systems is only possible with the aid of clever heuristics which significantly reduce the computational load.

Determining heuristic functions to guide chess play sufficiently well so as to challenge the best human players turns out to be extraordinarily difficult and computing these functions demands substantial computational resources. We will discuss this further in the following chapter when we contrast automated methods of game play with human methods.

Given a heuristic function, we still need to decide how to be guided by heuristic values in conducting a search. One such heuristic search procedure is *hill climbing*.

A hill climbing search involves evaluating, at each node, the heuristic value of its immediate descendant nodes and then moving to the node with the best heuristic value. If we reach a terminal node without finding a solution state, we backtrack to the next best possibility.

The letters in Figure 11.5 represent the order in which we would search the depicted generation tree if our heuristic assigned values to nodes as shown, where 0 represents a solution state.

Exercise 11.2

If we were to simply apply a breadth first, left first search to the tree depicted in Figure 11.5 (without applying a heuristic function), how many nodes would we have to traverse before reaching the same solution state? What if we used a depth first, left first search? What if we used depth first, right first?

As Exercise 11.2 demonstrates, this heuristic search is notably more efficient – in terms of the number of nodes searched – than either of our blind search procedures. It is also the case, though, that it is more computationally expensive *per node* since as well as determining

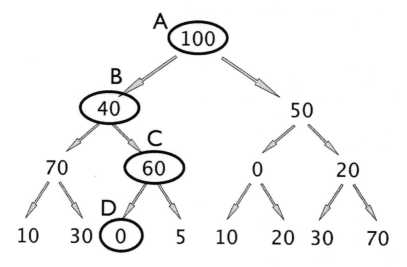

Figure 11.5   Hill climbing search.

whether each searched node represents a solution state, we are also applying a heuristic function.

For any generation tree of even moderate complexity though, the computational cost per node of applying the heuristic function will be far outweighed by the overall computational savings conferred by the heuristic guide, since we typically massively decrease the number of nodes we need to examine.

Unfortunately, however, heuristics are just that – heuristics. They are not decision procedures. In other words, while heuristic values provide us with some sort of indication of the closeness to solution of a given node, the assigned value may not be a good representation of the *actual* closeness to solution of the node. Better heuristics provide more accurate indications, but it is always the case that heuristic values can lead us astray.

One way in which hill climbing search can lead us astray is if a particular search path involves initially moving away from a solution state (according to the criteria measured by our heuristic) but then swings back and leads to a solution more quickly than other search paths.

To make this clear, consider the following spatial analogy. Suppose I am in the rainforest at an elevation of roughly 100 metres and I want to get down to the beach which I know to be roughly to my east. There are two paths I can see, one of which seems to head north-east and the other of which leads south-east. I can't see much of either path through the trees but what I can see is that the north-east path seems to lead gently

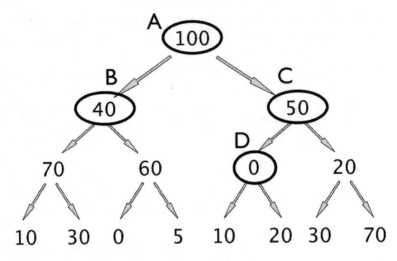

Figure 11.6   Best first search.

downwards, whereas the south-east path involves initially climbing a small rise. My goal is at sea level and I want to expend as little effort as possible so I take the north-east path. It turns out, however that the north-east path is meandering and involves climbing many rises along the way. The south-east path though, after cresting the initial rise, descends quickly to the beach, dropping in elevation all the way.

In the scenario described, a hill climbing search causes me to miss the quickest search path to the solution, as that path appeared initially disadvantageous. One way to accommodate such situations is to do a *best first* search.

A best first search is just like a hill climbing search save that our choice of nodes to move to is not restricted solely to immediate descendant nodes. We also keep a record of the heuristic values of *all* the nodes we have searched so far and we choose from among the collection of immediate descendent nodes and these other nodes.

This means that if, during a search, we reach a node where all the immediate descendant nodes are rated by our heuristic as further from our solution than a node we searched earlier, we can go back up the tree and search from the better rated node next.

As with hill climbing, best first search involves taking the path of least resistance. The difference is just that we expand our available choices from any given node – we no longer require that search paths always lead down a branch to the next iteration.

Figure 11.6 shows how the same tree from Figure 11.5 would be searched according to a best first search.

As we can see, using best first search allows us to discover, in this case, a shorter derivation for our solution state – we find a solution node at a prior iteration.

Best first search is clearly the most efficient search procedure we have examined. It should be clear, however, that it is also the most computationally expensive per node. Not only do we have to apply a heuristic function to nodes, but we also need the memory space in which to record the values of all the nodes we have searched so far. As the complexity of the tree increases, this computational expense can become significant. It is still the case, though, that the computational savings in terms of the number of nodes we need to search far outweigh the computational cost the heuristic procedure incurs.

There is a lot more for us to say about heuristics, so in the following chapter we are going to examine how me might employ heuristic search for a classical artificial intelligence application – playing games.

CHAPTER 12

GAMES

Now that we have a basic understanding of search procedures and heuristic functions, it is time to apply this understanding in consideration of automated methods of game play.

This provides us with an entry point to a broader examination of how we might employ formal systems to enact functions which are taken to be constitutive of intelligence.

We are also going to compare these formal methods for game play to our reflective understanding of how humans play such games, before moving on in the following chapters to a more detailed and informed such comparison with respect to the cognitive functions implicated in reasoning and language.

12.1 A SIMPLE GAME

If a game is sufficiently simple, we can very easily automate a procedure for playing it. It is fairly trivial, for instance, to determine an algorithm for playing tic-tac-toe such that following the algorithm will always result in either a win or a draw.

In such cases, all we need do is construct the generation tree for the game, then work backwards from the terminal nodes to determine a strategy.

For instance, suppose we have a game that is played as follows. First an initial position is chosen. From that position, and each subsequent position, there are exactly three ways in which a player might move. Players flip a coin to determine who begins, then the first player makes a move, the second player makes a move and the first player makes a final move. After these three moves, the game ends in a position which is a clear win for one of the players.

The generation tree for this game is easily constructed. It will have three iterations and a branching factor of three, so there will be

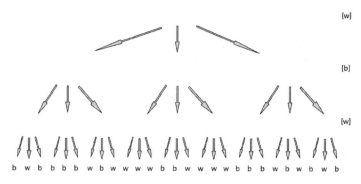

Figure 12.1   A simple game.

twenty-seven terminal nodes, each of which is discernible as a win for
one player (white) or the other (black).

The generation tree depicted in Figure 12.1 assumes we have
chosen an initial position for the game and assigns to each terminal
node the player for whom it is a winning position. We needn't know
the particulars of the game or its play – its form is all we're concerned
with here. I've merely stipulated which nodes are wins for which
player for illustrative purposes.

With the information depicted in Figure 12.1, we can work our way
back up the tree and determine who the chosen initial position will be
a win for.

Let's suppose, for instance, that white wins the toss and is to make
the first move. That means that on the last iteration before the terminal nodes, it is white's move. A node at that iteration will count as a
win for white – given that it is white's move – if there is at least one
descendant node which results in a win for white, otherwise it will
count as a win for black.

In other words, if white can move to a winning position and it is
white's turn, then the position is already a winning position for white.
If, however, the only available moves are to positions which are
winning positions for black, then the position is already a winning
position for black.

Similarly, at the previous iteration it will be black's turn. A node at
that iteration will count as a win for black if there is at least one
descendant node which counts as a win for black, and will count as a
win for white otherwise.

Figure 12.2 applies this procedure to the tree depicted in Figure
12.1 and demonstrates that if white wins the toss, white has a winning
strategy available. This is not, however, to say that white *will* win the

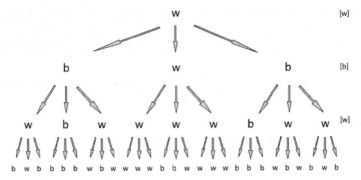

Figure 12.2 Winning strategy for white.

game – after all, a human player may well make a mistake. It is merely to say that there is an algorithmic *winning strategy* available to white.

Exercise 12.1

Assume that black wins the toss and complete the generation tree depicted in Figure 12.1 according to the procedure we have described. Is there a winning strategy available to black from this initial combination if black is to move first?

The solution to Exercise 12.1 demonstrates that for the game described, and given the initial position which results in the assignment of terminal nodes as depicted in Figure 12.1, the player who wins the toss has a winning strategy available to them. This may not be the case for other initial positions of the game but it is a deliberate feature of the example assignments of winning players to terminal nodes.

Exercise 12.2 (Challenge)

Consider the generation tree for tic-tac-toe. Assume that all nodes representing winning positions are terminal.

(a)   How many nodes are there at the first iteration?
(b)   At which iteration are terminal nodes first generated?
(c)   How many nodes are there on the tree in total?
(d)   How many terminal nodes does the tree contain?
(e)   How many of these nodes represent winning states for either player?

While this procedure of working backwards from terminal nodes is efficient in delivering winning strategies, its application is limited to only the simplest of games. As we saw in the last chapter, the generation trees for more interesting and complex games tend to grow exponentially, rendering a procedure such as this computationally untenable.

To determine methods of searching for winning strategies in games such as chess, we will need to employ heuristics.

12.2 MINIMAX

We saw in section 11.3 that we can apply a heuristic function to a node which evaluates, according to criteria relevant to the system, the closeness to solution of that node. In the context of a two-player game, the states which will count as solution states are those representing wins for (an arbitrary) one of the players. In this case, the furthest a game could be from a solution state would be a state representing a win for the other player.

Consequently, our heuristic values will represent how a game state stands with respect to a possible win for either player. For the sake of example, lets stipulate that lower numbers will represent closeness to a win for black and higher numbers will represent closeness to a win for white.

So unlike the example we saw in section 11.3 – in which the root node was as far from solution as possible – the root node for a two-player game will take a heuristic value exactly in the middle of the range (presuming the initial position is not prejudiced towards either player).

Heuristic functions evaluate nodes based on certain internal features of the state represented at the node. So, for example, in chess we can take account of material advantage, dominance of the centre and advancement of certain pieces in order to generate a value which represents the goodness of that state for each player. Even a very good heuristic function, however, is limited in only considering features *internal* to the state.

Contrast this with the procedure we considered in the previous section. Using information about states generated at further iterations to evaluate prior nodes is a procedure which considers features *external* to states.

We can greatly increase the accuracy of a method for determining strategies in complex games by using a combination of internal and external features as a guide. In other words, we can combine the use of a heuristic function with a method for working backwards from evaluated nodes to determine a value for prior nodes.

We've seen that we can't feasibly construct the entire generation tree for chess. What we can do, however, is search ahead just a few moves and apply a heuristic function to the nodes which are at the *search horizon*. We can then use a *minimax* procedure to work backwards from the search horizon to determine a value for the node we're evaluating.

Given our earlier stipulation that lower heuristic values will represent closeness to a win for black, black will be a *minimiser* and white will be a *maximiser*. In other words, black will always be seeking to move to states which have lower heuristic values, and white will seek to move to states with higher values.

Let's make a minor modification to the simple game we described in the previous section, such that the game no longer always results in a clear win for one player at the end of three turns. Now let's suppose that we select an initial position and draw up the generation tree for *only* the first three moves, then apply a heuristic function to get the values depicted at the *horizon nodes* in Figure 12.3.

Using the heuristic values at the horizon nodes, we can apply a *minimax* procedure to determine a value for the selected initial position, as follows.

Assume white wins the toss, in which case at the iteration immediately prior to the search horizon it is white's turn to move. White is a *maximiser* so the value of a node at that iteration will be the *maximum* of the values of the immediate descendant nodes.

At the iteration immediately prior to that, it is black's turn to move. Black is a *minimiser* so the value of a node at that iteration will be the *minimum* of the values of the immediate descendant nodes.

Using this procedure, we work backwards from the horizon nodes to determine the value of the root node, as Figure 12.4 depicts.

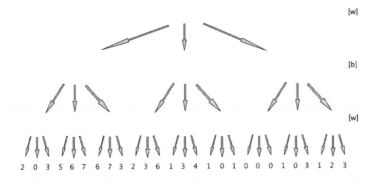

Figure 12.3 Minimax.

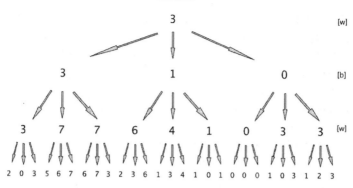

Figure 12.4 Winning strategy for white.

Exercise 12.3

Assume that black wins the toss and – using the same values for horizon nodes depicted in Figure 12.3 – employ a minimax procedure to determine the value of the initial position.

12.3 PRUNING

We now have the makings of a procedure for searching for strategies for playing a game as complex as chess. We search as far ahead as is computationally tenable, apply a heuristic function to the nodes at the search horizon, then use a minimax procedure to work back to the node we are searching from.

This procedure, however, still suffers from combinatorial explosion. Fortunately, there are further ways to maximise the efficiency of our computational resources by cutting down the amount of searching we have to do, or *pruning* the search tree.

One such method involves using what we have already discovered in a search to determine that we can disregard certain branches. The tree fragment depicted in Figure 12.5 shows a (greatly simplified) situation in which this is possible.

Given that black is a minimiser, we know that the value of node *b* will be 3. We also know that the value of node *c* will be *at most* 2. Consequently, we already know that the value of node *a* will be 3 (since white is a maximiser) regardless of the value of the node marked with a question mark.

Certainly this is a trivial pruning in the example case. However, it should be clear that the principle will extend to significantly more

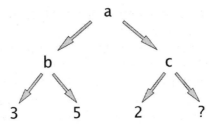

Figure 12.5   Pruning the tree.

complex cases and may allow us to disregard large parts of a search tree, based on what we have antecedently discovered.

Another method for pruning is to find guiding principles which tell us to simply not bother considering certain moves. It is often the case with complex games that there will be certain legal moves available from a given state that no one playing well would actually make, or that no one would be at all likely to make. If we are informed with respect to the unlikelihood of these moves obtaining, we can disregard the branches of the search tree descending from those nodes.

A further pruning method involves playing out certain sequences of moves from memory, without engaging in any search. In fact, this is precisely how computers typically play chess openings. They have access to a large database of standard openings and variations and play the initial moves from a predetermined sequence.

The more we are able to prune the search tree, the more effectively our computational resources will be deployed. Consequently, in order to program a machine to play a complex game well, we will need to employ the best heuristics we can devise, in conjunction with various methods for simplifying the search task.

12.4 HUMANS VERSUS COMPUTERS

We have now given some consideration to an activity which is generally taken as constitutive of intelligence – the ability to (at least learn to) play well a complex game such as chess. We've seen – at least in broad strokes – how to employ symbol systems and search methods to automate the strategic play of such games.

In the case of chess, this involves a large database of opening strategies, an optimised heuristic function for evaluating game states, various principles for disregarding unlikely moves, and the deployment of significant computational resources to evaluate a very large number of states per second. As I write this, the machines which are used to challenge chess grandmasters are able to evaluate around two hundred million states per second.

Let's compare what we've learned about automated chess play to what we intuitively know about human chess play.

For starters, humans – at least those who play very well – also memorise large numbers of opening strategies and variations thereof. For seconds, humans are also very good at determining which moves are unlikely to be made by an opponent playing well. Those humans who are very experienced at chess also have a very good sense of how good a particular game state is for them.

It seems then that there is a good fit between what we know of automated chess play and what we can intuit concerning the play of accomplished humans. There are, however, important distinctions between the two, the most obvious of these being the sheer quantity of search and evaluation carried out by chess computers.

It is clear that chess grandmasters do not *explicitly* apply a heuristic function to some two hundred million states per second in order to evaluate the goodness of the current game state and determine a strategy. We need to be very careful here, however, as we might be tempted to countenance a very poor argument against computationalism in light of this.

The very poor argument runs as follows. Computers need to search two hundred million states per second in order to match the best human players. These human players, however, clearly do not. So human players are not enacting formal systems in playing chess. Therefore computationalism is false, since there is at least one cognitive task whose function cannot be accounted for in terms of the operations of formal systems.

There are a number of reasons why this argument is bad. The most telling criticism of it is that there is a distinction between human performance and our best attempts at recreating it.

It may well be the case – and will be if computationalism is true – that human players do actually employ these methods. It is just that humans have very good guiding principles that allow them to engage in significantly less search to determine winning strategies. The fact that these human players are unable to make *explicit* the principles they are guided by does not speak against the possibility that their cognitive functions are to be accounted for in terms of the operations of formal systems. After all, much of our mental life is opaque to introspection. It merely speaks against the best results of our empirical investigation in seeking to determine the operations of these systems.

Another criticism of the argument against computationalism is that it equivocates between the cognitive functions we explicitly and consciously engage in and cognitive function per se. To say that a grandmaster does not *explicitly* consider two hundred million states per second is not equivalent to saying that their cognitive processes of evaluation are not implicitly tantamount to consideration of such a number of states. Relying on a claim of the latter when only the former is demonstrable is an equivocation which begs the question against the computationalist.

While we clearly need to be careful about the conclusions we draw from perceived distinctions between computer chess play and human chess play, there are yet further differences that are worth mentioning.

For one thing, the minimax procedure assumes that one's opponent will always make the move which is optimal for them. Very often, however, it will be the case that a human player will make a sub-optimal move. It may be the case that they fail to recognise the optimal move, or it may be the case that they deliberately play a sub-optimal move to try to force their opponent into following a particularly strategy.

A computer may well, of course, also make a sub-optimal move, simply through failure to correctly evaluate. Deliberately making a sub-optimal move to achieve an effect in one's opponent's strategy is, however, another thing entirely.

In order to actively seek to influence the strategy of one's opponent by making an unusual move, one requires an understanding of how one's opponent thinks about the game. In other words, one requires some appreciation of how one's opponent is likely to move given certain circumstances. The best way to discern this is to play as much as possible against the opponent in question.

During computer versus human tournament play, it is generally allowed for the programmers to modify the computer program between games in order to counter strategies deployed by the human player. It is important to note that where the computer requires human intervention to be able to respond to the idiosyncrasies of a given player's strategy, rapid responsive adaptiveness to peculiarity seems to be something that humans naturally excel at. This leads us to the final point of distinction between human and computer game play we'll consider here.

Consider the fact that the principles which guide human evaluation of chess states are sufficiently superior to those employed by a computer as to require significantly less explicit evaluation of descendant states in order to play comparably well. Consider also the rapid responsiveness to the idiosyncrasies of a given player's strategy which is evidenced by human players but not computers.

It appears that there is something subserving both these abilities which humans are very good at but which hasn't been mentioned at all in this discussion of machine game play. Humans excel at *extracting patterns* from environmental stimulus and recognising when these patterns obtain again, even when they only partially obtain, or when there is some variation in the pattern.

This is an important insight and one which we will return to discuss at great length in Chapter 19. For the moment, note that this is not *ipso facto* an argument against computationalism, but it does place an explanatory burden on the computationalist – the burden being to give a computational account of this cognitive capacity.

Now that we have compared, at least in rudimentary fashion, the way humans play complex games with the way computers can be programmed to play these games, it is time to turn our attention to the human rational and linguistic capacities.

CHAPTER 13

MACHINE REASONING

In this chapter we are going to begin to investigate the rational capacity – the ability to *reason*. A thorough survey of automated reasoning methods would require dedicated volumes. We're going to concentrate on one kind of automated reasoning project which suits our purposes well – the design of *expert systems*.

We're going to see how we might recreate, with the use of formal systems, the reasoning processes of a human expert in a particular domain.

Before we do so, we will first need to make clear some concepts and terminology involved in the study of logic.

13.1 LOGIC AND DEDUCTION

The first thing to appreciate is that there is a distinction between logic and logics. Logic is a research tradition whose objects of investigation are logics. These logics are formal systems and there are very many of them.

The aim of logics is to formally encode relations of *entailment* or *logical consequence*. In other words, logics are formal systems which provide methods for determining what *follows* from what as a matter of *logical form*.

Another important point is that, like all formal systems, logics are concerned only with formal properties. While it is the case that elements of logics are *interpretable* as meaning something, issues of meaning never have any bearing on determinations of logical consequences. In other words, whether or not something follows logically from some other things is *entirely* a question of their respective logical forms. The relevance of this will become apparent in exercises later in this chapter.

I don't expect you to yet have an understanding of precisely what logical form is or how to discern logical forms. I don't intend this to be a fully fledged introduction to logic so I'll reserve further discussion concerning this for the next section where we'll see some simple examples.

Given that logics are formal systems, you should be wondering what the states are and what the rules are. For our purposes, states can be thought of as sets of statements. Rules of logics are such that given an input state containing some statements of a certain form, we can derive new statements of a certain form to add to the output state.

This process of applying logical rules to sets of statements to generate novel statements is the process of *deduction*.

While we will be concentrating on deduction in this chapter, it is important to realise that there are other distinct kinds of reasoning. We'll revisit this issue at length in Chapter 15 and again in Chapter 19 but for the moment let's just briefly consider the distinction between deduction and *induction*.

Induction is the form of reasoning employed by empirical science. Inductive proof is rather a different thing from deductive proof. A deductive proof is a demonstration – of the kind we will see in the following section – that certain statement forms can be derived from other statement forms according to certain logical rules. A typical *inductive* proof, on the other hand, is a display of amassed observations, made under certain conditions, in support of the claim that future such conditions will yield the same observations.

It is a feature of deductive proofs that they are unrevisable. There is nothing that can be added to a deductive proof such that it will, in light of the addition, fail to yield its original conclusion. Inductive proof, on the other hand, is essentially revisable. Inductive proofs only establish their conclusions with a certain degree of probability and are only ever one countervailing observation away from failing to deliver their purported conclusions.

Although I've only given a rough sketch of inductive reasoning here, I've illustrated the distinction with deductive reasoning for two reasons: firstly, in order that you appreciate that there are legitimate and established kinds of reasoning which are distinct from deduction; secondly, so that you realise that the word 'proof' means something quite different in the mouths of scientists to what it does in the mouths of logicians or mathematicians. When you hear that scientists have 'proven' something or that 'studies have shown' something, it pays to realise that the very thing proved or shown may be revised and disproved in light of subsequent investigation.

There is a lot more to be said about scientific reasoning and I certainly wouldn't want to be charged with having given only a caricature of scientific process so, once again I refer you to the suggestions for further reading.

Before we proceed to examine expert systems, we will need to develop a little bit of terminology concerning *conditionals* and *predicates*.

13.2 CONDITIONALITY AND PREDICATION

Natural language conditionals are statements of the form 'if . . . then . . .'. The study of conditionals, and the determination of an adequate formal account thereof, is of central importance to logic. Many logics are distinguished solely by virtue of their treatment of the conditional.

We can represent conditionals by using an arrow. The statement 'if today is Monday then tomorrow is Tuesday' can be represented as follows:

today is Monday $\rightarrow$ tomorrow is Tuesday

The left-hand side of a conditional – which represents the bit between 'if' and 'then' – is the *antecedent* of the conditional. The right-hand side – the bit which comes after 'then' – is the *consequent* of the conditional.

If the antecedent of a conditional is *satisfied* then we can derive the consequent according to a simple logical principle. So, if it is actually the case that today is Monday, we can – given the above conditional – *deduce* that tomorrow is Tuesday. This logical principle is known as *modus ponens* and can be symbolised as follows:

$$\Phi \rightarrow \Psi$$
$$\Phi$$
$$\overline{\qquad\qquad}$$
$$\therefore \Psi$$

The logical principle of *modus ponens* – which tells us that given a conditional with a satisfied antecedent we can deduce its consequent – is the only logical principle we will be appealing to in our examination of expert systems.

The last thing to do before looking at an example expert system is to discuss *predicates* and logical forms.

Consider the following two statements. If something is a dog then it is a mammal. If something is a mammal then it has a heart. One way to represent these statements would be as follows:

something is a dog $\rightarrow$ that thing is a mammal
something is a mammal $\rightarrow$ that thing has a heart

However, we can do better than that. Notice that in each antecedent and consequent, a property is applied to – or *predicated of* – a thing. Note also that in each conditional, it is the *same thing* referred to in both the antecedent and the consequent.

If we take 'dog' to represent the property of being a dog, 'mammal' to represent the property of being a mammal, and 'heart' to represent the property of having a heart, we can recast the above conditionals to capture the fact they are applying properties to the same thing in their antecedents and consequents:

dog $(x) \rightarrow$ mammal (x)
mammal $(x) \rightarrow$ heart (x)

These conditionals are as close to logical form as we require for the purposes of this chapter. The symbols 'dog', 'mammal' and 'heart' – which could, of course, be substituted uniformly for any other symbol we choose – represent *predicates*. For our purposes, predicates can be understood as encoding properties and relations.

The symbol x in the above conditionals is a variable – as you have no doubt discerned. We will say that the antecedent of one of these conditionals is *satisfied* if we have a statement which has the same *logical form*.

Statements, for our purposes, apply predicates to *names* (not variables). So if, for instance, we know that Mia is a dog, we can represent this by using the symbol m as a name for Mia, as follows:

dog (m).

We can now use the two conditionals we have symbolically represented to do some simple deduction. The statement – dog (m) – is of the same logical form as the antecedent of our first conditional. This means that the antecedent of the conditional is *satisfied* so we can *deduce* the consequent, namely mammal (m). We now have a statement which satisfies the antecedent of the second conditional, so we can deduce its consequent and derive heart (m). Given that we know that 'heart' represents the property of having a heart and that m is a name for Mia, we have just deduced that Mia has a heart.

We can represent this deduction symbolically, as follows:

dog (m)
dog $(x) \rightarrow$ mammal (x)

$\therefore$ mammal (m)

mammal $(x) \rightarrow$ heart (x)

∴ heart (m)

Exercise 13.1

(a) If something is a woman, then that thing is human. If something is human, then that thing is mortal. Represent this symbolically.

(b) Sue is a woman. Give a symbolic representation of the deduction you can make from this, given the conditionals you represented in (a).

The predicates we have seen so far have been *one-place* or *monadic* predicates. We will also want to use, in our example expert system, predicates which attribute relations between two names.

Consider the relation expressed in the statement 'Mia is older than Linus'. The relation 'older than' is asserted to hold between two things which we can name m and l, allowing the following symbolic representation:

older_than (m , l)

Note that the order of names in two-place predicates is important. The following statement is interpreted as saying that Linus is older than Mia.

older_than (l , m)

We can use the symbolism we have developed so far to encode something that we know – as a matter of common-sense knowledge – about the relation 'older than', as follows:

older_than (x , y) & older_than $(y , z) \rightarrow$ older_than (x , z)

The ampersand (&) in the above simply stands, as you might expect, for 'and', the logical operation of *conjunction*. We say that the conditional represented above has a *conjunctive* antecedent. The symbols x, y and z are variables, not names.

The above conditional tells us that if x is older than y and y is older than z, then x is older than z. This is something that anyone who understands the meaning of 'older than' implicitly understands. In logical terms, it tells us that the symbol 'older_than' represents a *transitive* relation.

Exercise 13.2

(a) What other relations can you think of which are transitive?
(b) A *symmetrical* relation is one such that if x bears the relation to y then y also bears the relation to x. What symmetrical relations can you think of?
(c) Use the symbolism we have developed to represent that a particular relation is symmetrical.

If we know that Mia is older than Linus and we also know that Sue is older than Mia, then using s as a name for Sue, we can reason as follows:

older_than (s , m)
older_than (m , l)
older_than (x , y) & older_than $(y , z) \rightarrow$ older_than (x , z)

———————————————————————————————

∴ older_than (s , l)

The conjunctive antecedent of the conditional is satisfied because both *conjuncts* – the statements flanking the ampersand – are satisfied, so we have licence to deduce the conclusion by *modus ponens*, as before. I'm sure you are quite able to discern what it is we have proven.

This is all the terminology and symbolism we require to develop an expert system.

13.3 KINSHIP

If you were able to follow the example deductions in the previous section, then you already grasp the important aspects of the operations of expert systems. In fact, the example cases used to introduce predicate notation were actually themselves miniscule expert systems.

Expert systems are formal systems which aim to encode the information that a relevant human expert knows about a particular domain of knowledge and to reproduce their deductive processes given this information and some novel input. Our example expert system is going to encode information concerning kinship relations.

The *resident information* of an expert system is specified in terms of a number of conditionals. This resident information serves as the rules of the expert system.

While we will be appealing to a logical principle – *modus ponens* –

in *applying* the rules to make deductions, the rules themselves are not rules of logic. Rather, the rules of the system are conditionals which represent the information an expert knows about the relevant domain of knowledge – in this case kinship. The logical principle will be captured in the effective procedure for applying the rules.

The resident information of our kinship system is specified as follows:

grandparent_of (x, y) & male $(x) \rightarrow$ grandfather (x)
parent (x) & male $(x) \rightarrow$ father (x)
parent (x) & female $(x) \rightarrow$ mother (x)
parent_of $(x, y) \rightarrow$ parent (x)
parent_of $(x, y) \rightarrow$ child_of (y, x)
parent_of (x, y) & parent_of $(y, z) \rightarrow$ grandparent_of (x, z)

The states of an expert system are sets of statements. The initial state of our kinship system will be:

parent_of (j, m)
parent_of (m, h)
male (j)
male (h)
female (m)

The final thing to specify for our kinship system is the effective procedure for applying rules to states.

1. Starting with the first conditional in the resident information, check to see if there is a statement in the state which satisfies the antecedent – i.e. check to see if any of the statements in the state have the same logical form as the antecedent and differ from it only in substituting the variable(s) for name(s).
2. If there is a statement which satisfies the antecedent, then add the consequent to a list of deduced statements (being careful to substitute the variable(s) in the consequent for the same name(s) as those in the statement(s) which satisfied the antecedent). Check for further statements in the state which satisfy the antecedent.
3. Repeat steps 1 and 2 for each conditional in the resident information. When this is completed, augment the original state with the list of deduced statements and output this augmented state. Only add a statement from the deduced list if it does not already appear in the state.
4. Begin again with the first conditional and see if the new statements deduced allow the deduction of further novel statements.

5. If a state is such that none of the conditionals in the resident information allow the deduction of statements that are not already in
   the state, then halt.

Let's apply this procedure to our initial state and see what we can
derive.

Exercise 13.3

Before reading on, attempt to apply the first three steps of
this procedure to the initial state of our kinship system.

The first conditional in our resident information is:

grandparent_of (x, y) & male $(x) \rightarrow$ grandfather (x)

We don't have any statements in our initial state of the form grand-
parent_of (x, y), so we are unable to satisfy the antecedent of this
conditional. We can satisfy one of the conjuncts as we do have statements of the form male (x), but we need to satisfy *both* conjuncts to
satisfy the conjunctive antecedent.

The next two conditionals in our resident information are:

parent (x) & male $(x) \rightarrow$ father (x)
parent (x) & female $(x) \rightarrow$ mother (x)

Once again, we don't have any statements of the form parent (x), so
we can't satisfy the antecedents of either of these conditionals.

Do be careful here, for while we do know that parent_of (j, m),
and we know – as a matter of common sense – that if someone is a
parent of someone else they are *ipso facto* a parent, we can't just
assume 'parent (j)'. The symbols 'parent' and 'parent_of' are just
that – symbols. We have used symbols that are meaningful to us but
as far as the operations of the system are concerned, any relations
between the predicates that symbols represent need to be encoded
explicitly in the rules. This is precisely what the next conditional
does.

The fourth conditional in our resident information is:

parent_of $(x, y) \rightarrow$ parent (x)

We do have a statement in the initial state which satisfies the
antecedent of this conditional – parent_of (j, m) – so we can add the
consequent (being careful to substitute the correct name for the variable) to our list of deduced statements – namely parent (j).

We also have another statement in the initial state which satisfies the antecedent of the same conditional – parent_of (m, h) – so we can also add parent (m) to our list of deduced statements.

The fifth conditional in our resident information is

parent_of $(x, y) \rightarrow$ child_of (y, x)

We have two statements in the initial state which satisfy the antecedent of this conditional – parent_of (j, m) and parent_of (m, h) – so we can add child_of (m, j) and child_of (h, m) to our list of deduced statements.

The final conditional in our resident information is:

parent_of (x, y) & parent_of $(y, z) \rightarrow$ grandparent_of (x, z)

We need to be careful here – the name which we substitute for y in the first conjunct of the antecedent must be the same name which we substitute for y in the second conjunct. As it turns out, we do have statements which satisfy the antecedent of this conditional – parent_of (j, m) and parent_of (m, h) – which allows us to deduce grandparent_of (j, h).

We have now considered each of the conditionals in the resident information and have deduced five new statements:

parent (j)
parent (m)
child_of (m, j)
child_of (h, m)
grandparent_of (j, h)

So we add the list of deduced statements to our initial state (checking to make sure none of them are redundant) to get the following state:

parent_of (j, m)
parent_of (m, h)
male (j)
male (h)
female (m)
parent (j)
parent (m)
child_of (m, j)
child_of (h, m)
grandparent_of (j, h)

The next thing to do is to check each of the conditionals again in turn to see if the new statements in our derived state allow us to deduce any further novel statements.

Exercise 13.4

Before reading on, determine what new statements, if any, can be deduced on the second pass through the conditionals in the resident information. What about the third pass?

The second pass through the conditionals in the resident information will allow us to deduce three novel statements:

grandfather (j)
father (j)
mother (m)

A third pass generates only redundant statements – statements which are already in our generated state – so we halt.

Exercise 13.5 (Challenge)

Augment the initial state of our kinship system with the statements:

parent_of (m , j)
female (j)
diff (j , h)
diff (m , h)

and add the following conditionals to the resident information:

parent_of (x , y) & parent_of (x , z) & diff (y , z) → siblings (y , z)
siblings (x , y) & male (x) → brother (x)
siblings (x , y) & female (x) → sister (x)
siblings (x , y) → siblings (y , x)

Apply the rules to generate all the statements you can. Do things appear a little strange? Why?

13.4 EXPERT SYSTEMS

The example expert system of the previous section is greatly simplified but, nonetheless, it is clear that it is able to capture the deductive process that we engage in when reasoning about kinship relations.

It may seem cumbersome and artificial compared to our thought processes but this is only because the kind of reasoning we do when told that someone is a female parent is rapid and automatic. We don't need to *explicitly* apply a rule to determine that the person in question is a mother – this is simply something we automatically understand when we understand that they are a female parent. This doesn't, however, speak against the claim that this implicit understanding is governed by just such methods.

In fact, if you are asked to determine what relation to you your mother's sister's daughter's husband is, you are likely to have to think more explicitly about the kinship relations involved, following rules very much like the ones we encoded in the previous section.

Interesting expert systems are, of course, considerably more complex than our example. Our kinship system has very little resident information and appeals to only one logical principle – *modus ponens*. More complicated expert systems will involve considerably more resident information and will appeal to numerous logical principles in applying rules.

Consequently, we will often want to generate a bottom-up search to determine whether or not a particular statement is included in any generated state. In such cases we begin with just the statement we are interested in deriving and work backwards through the rules, considering not entire states but, rather, just those statements which must be included in a state in order for us to have derived the statement(s) at the previous iteration. If a statement at a node is included in the initial state, then we can strike it off at the next iteration. If we get to a node where nothing needs to be included in the state to prove the statement at the previous iteration – i.e. the statement(s) we did need to generate the state we are interested in have been shown to be included in the initial state – then the search ends in success.

Figure 13.1 demonstrates a bottom-up search for the state father (j) in our kinship system. At the first iteration, we work backwards through the available rules to determine that the only way to generate father (j) is if we have parent (j) and male (j).

At the second iteration, there are three branches, representing the three statements which would allow us to generate parent (j). The left-hand and right-hand branches end in failure since neither parent_of (j, j) nor parent_of (j, h) are in the initial state and there is no rule which allows us to generate parent_of statements.

The middle branch continues since we have parent_of (j, m) in the initial state so we discharge this statement at the next descendant node, leaving only male (j) to prove. At the next iteration we discharge

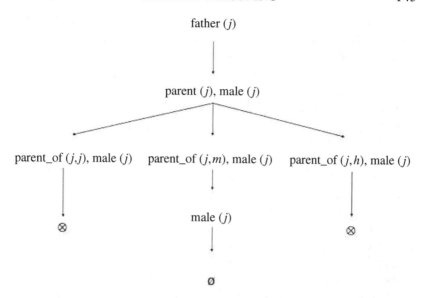

Figure 13.1   Bottom-up search.

this last statement as it too is in our initial state, leaving nothing (the empty set) to prove. So a search path down the middle branch ends in success and we read the derivation off by following the branch back up to the root node.

Exercise 13.6

Generate a bottom-up search in our kinship system for the states:
(a)   mother (m);
(b)   grandfather (j).

Expert systems can be usefully deployed in weak artificial intelligence projects. One such well known system is MYCIN which was developed in the 1970s at Stanford. MYCIN can make probabilistic diagnoses of pathologies and recommend medication based on the results of blood tests. Its resident information encodes heuristic procedures followed by medical doctors in making rough and ready diagnoses in the absence of developed cultures. Its rule application procedure appeals to probabilistic reasoning principles so it is able to suggest a number of possible diagnoses with a certainty factor attached to each.

More interestingly for our purposes, there are some who hold out hope that expert systems alone can give rise to strong artificial intelligence. The Cyc project, founded by Douglas Lenat and developed under the auspices of the private research institution Cycorp, aims at precisely this. The conviction held by Cyc researchers is that if they can encode – in the resident information of the system – all (or much of) the information that you and I take to be common knowledge, and develop a sufficiently sophisticated inference engine for making deductions, they will thereby develop a strong artificial intelligence artefact.

Whether or not expert systems methods are sufficient for artificial intelligence remains an open empirical question. There are reasons, however, to believe that there may be problems with this approach. Some of these we will consider in Chapter 15 and some in Chapter 19. In defence of the Cyc project, however, it appears – from what I can gather of their operations – that they are responsive to at least some of these concerns and are augmenting the traditional expert system model accordingly.

Perhaps the most significant and difficult computational problem which needs to be solved on the way to true artificial intelligence is the problem of interpreting and producing natural language. It is to this issue that we turn our attention in the next chapter.

CHAPTER 14

MACHINES AND LANGUAGE

Devising computational procedures for handling natural language is arguably the most significant problem facing artificial intelligence researchers. In this chapter we're going to begin by considering the various computational problems which need to be solved in order to facilitate this.

As is the case with machine reasoning, a comprehensive survey of computational methods for facilitating linguistic interpretation and production would require dedicated volumes. We're going to concentrate here on just one of these methods, in service of one of the functions constitutive of linguistic competence – determining grammaticality.

We will lend further consideration to the procedures governing linguistic activity in Chapter 16 and will return to examine computational methods for implementing further functions subserving the linguistic facility in Chapter 19.

14.1 INTERPRETING LANGUAGE

Let's reflect on the various procedures involved in the comprehension of a spoken utterance.

Spoken language is generally delivered in a continuous phonetic stream which does not readily reveal its linguistic properties. While it might seem that individual words are easily discernible from phonetic properties of an utterance alone, this is generally not the case unless one is speaking – very – slowly – and – carefully. This is clear if you look at a visualisation of a waveform of a recorded utterance. Furthermore, the *absolute* phonetic properties of an utterance – such as pitch and volume – will differ significantly from speaker to speaker. Interpreting a spoken utterance is a non-trivial function – in fact it is quite a complex procedure.

There are several tasks that mediate the interpretation of auditory input as a sentence of natural language. One of these tasks involves converting the phonetic input to a *phonemic* representation.

Phonemes are the atomic meaningful speech sounds of which a spoken language is constituted. We will learn a lot more about phonemes in Chapter 16 but for now a rough description will serve.

Phonemes are idealisations to which actual speech sounds – *phones* – approximate and which represent *distinctive contrasts* in a particular language. The initial sound in the word 'pat' and the initial sound in the word 'bat' are two phonemes which differ only in terms of *voicing*. The phoneme /p/ is not voiced but /b/ is. This means that when you produce the phoneme /b/ your vocal chords vibrate but when you produce /p/ air merely passes noiselessly through. Try putting your fingers on your voice box as you pronounce 'pat' and 'bat' and you should be able to tell the difference in voicing. We can tell from the fact that 'pat' and 'bat' *mean* different things that voicing is a *distinctive contrast* in English and that /p/ and /b/ are distinct phonemes.

This is a very rough and ready account of phonemes but for present purposes we merely need to appreciate that one of the mediating tasks in interpreting a spoken utterance involves determining, from the phonetic properties of the utterance, a representation of the distinctively contrasting sound units which are meaningful in the language uttered.

Another mediating task involves parsing the phonemic representation into a syntactic structure. In other words, we need to determine how the string of meaningful speech sounds breaks up into words, phrases, clauses and sentences. Finally, we need to work out what these words, phrases and clauses *mean*.

So in all, there are three distinct levels of representation implicated in understanding an utterance – phonemic representations, syntactic representations and *semantic* representations.

Producing an utterance involves these same three representational transformations. We intend a particular meaning, determine the syntactic structure which encodes that meaning, convert the syntactic representation to a phonemic representation and finally produce phonetic output for each phoneme. This production of phonetic output is governed by the phonemic environment in which a phoneme is situated – a feature of language production we will examine in Chapter 16.

Although the sequential manner in which I've enumerated these transformational stages might lead you to think that each stage of processing occurs separately and in sequence, there is good evidence to suppose that all three occur in concert.

For instance, it seems that the determination of syntactic structure is influenced by expectations governed by the semantic representation we are constructing as we syntactically parse an utterance. To make this clear, consider the following three sentences.

[1]   The horse raced past the barn.
[2]   The horse raced past the barn yesterday.
[3]   The horse raced past the barn fell.

Sentence [1] fits the prototypical syntactic pattern of an English sentence – it is composed of a noun phrase followed immediately by a verb phrase which consists of a transitive verb and its complement. Sentence [2] also fits this pattern – it merely adds an adjunct after the verb phrase.

Sentence [3], on the other hand, presents initially as unusual and potentially ungrammatical. We hear a noun phrase 'the horse' followed immediately by what we take to be the main verb of the sentence – 'raced' – and expect the remainder of the sentence (or at least the clause) to consist of the complement of the verb (past the barn / in the three o'clock at Flemington) and possibly one or more adjuncts (rather quickly / last Tuesday).

Consequently, when 'past the barn' is followed by the verb 'fell' we are taken aback as we were expecting maybe an adjunct but not another verb. We had already determined a semantic representation for the sentence up until that point but the presence of the final verb shows us that either we were wrong about the meaning (in thinking it was the meaning of [1]) or that the sentence is ungrammatical.

In other words, the meaning we construct as we interpret the sentence – in combination with what we implicitly know about the syntax of typical English sentences – conditions our expectations of syntactic structure. We assume the most likely syntactic structure and construct our semantic representation accordingly, which then influences our expectations of the remaining syntax.

It can, in fact, be quite difficult to get past the seeming ungrammaticality of [3] until we realise that the sentence doesn't have the typical syntactic structure we initially assumed. The verb 'fell' is the main verb of [3] and 'raced past the barn' merely indicates *which* horse it was that fell – the one which was raced past the barn. Note that we can substitute 'painted purple stripes' for 'raced past the barn' to achieve a similar effect, although the effect will be weaker as we are less likely to assume the horse to be the painting agent than the racing agent. If, however, we substitute 'running quickly' or 'belonging to Anne' for 'raced past the barn', the sentence becomes unambiguous.

Sentences with ambiguous syntax provide us with evidence that the construction of semantic representations and syntactic representations occur in concert and affect each other. Homophonic phrases provide further evidence that parsing phonemic representations into syntactic elements is influenced by both syntactic and semantic expectations.

Homophones are sequences of phonemes that have more than one semantic interpretation, such as 'pair' and 'pear'. A special case of homophony is homonymy. Homonyms are homophones which also have the same orthographic representation – i.e. they're written the same way, such as the verb 'bank' and the noun 'bank'. There are very many homonyms in English as many words can be interpreted as nouns or as verbs – e.g. paint, spring, void, power, urge, whisper, sleep, etc.

Homophonic phrases are sequences of phonemes which can be parsed as more than one distinct syntactic structure. An example homophonic phrase is the sequence of phonemes which can be interpreted as 'way up high' or 'weigh a pie'.

We can interpret a line sung by Judy Garland in *The Wizard Of Oz* as either of the following two sentences:

[4]   Somewhere, over the rainbow, way up high.
[5]   Somewhere, over the rainbow, weigh a pie.

Both of these are perfectly grammatical syntactic constructions – [4] is a description and [5] is an imperative. Consequently, syntactic considerations alone are not sufficient, in this case, to disambiguate between the two possible interpretations.

Our semantic interpretation, however, inclines us strongly towards [4]. The semantic interpretation of 'Somewhere, over the rainbow' makes it a much more likely hypothesis that Judy Garland goes on to further describe this place and *ipso facto* a much less likely hypothesis that she is issuing a pie weighing imperative.

In cases such as the following, however, it is not clear if it is syntactic or semantic considerations driving the disambiguation of the homophonic phrase:

[6]   I went the baker's to weigh a pie
[7]   * I went to the baker's to way up high

Sentence [7] is ungrammatical (as the asterisk indicates) so, in this instance, syntax alone is sufficient to disambiguate the homophone; however, it is difficult to separate the syntactic consideration from the influence of the semantic association. Fortunately, this is not something we have to rule on here.

What I want to concentrate on in the remainder of this chapter is the fact that we readily and automatically recognise [7] as an ungrammatical sentence. I want to examine how we might employ symbol systems methods to rule on the grammaticality of strings of written language.

14.2 GENERATIVE GRAMMAR

Noam Chomsky revolutionised the discipline of linguistics in the 1950s by taking a new approach to the study of grammar.

Grammar, before Chomsky, involved little more than taxonomising parts of speech and enumerating prescriptive principles that students of grammar should abide by. For instance, students of the prescriptive grammarians were told to never split an infinitive.

Chomsky, in contrast, took the grammar of a language to be the mechanism by which all and only the grammatical strings of the language can be *generated*. Chomsky argued that our internalised knowledge of the systematic *generative grammar* of our language accounts for the infinite *productivity* of language.

Language is infinitely productive in that we are able to produce, and rule on the grammaticality of, an infinite number of sentences, despite only ever having been exposed to a finite number. As a native speaker, you can immediately tell that the sentence 'Michelle's grandmother sells drugs to bikers in Belarus' is a grammatical sentence of English, even though it is unlikely that you've ever encountered that particular sentence before.

Formal systems are prime candidates for the mechanisms which facilitate the productivity of language. We've seen how formal systems can recursively generate an infinite number of states from finite resources in a rule governed fashion. We've also seen how we can conduct a bottom-up search to determine whether or not a particular state is generated in a given system. Now we're going to see how we might use formal systems to specify the *generative grammar* of a language.

The generative grammar for a language is a particular kind of formal system. It is a symbol system similar to the systems [STR] and [BIN] from Chapter 7. Some of its symbols – those which will appear at terminal nodes of its generation tree – can be interpreted as lexical items (words) of the language. The rest of its symbols can be interpreted as grammatical categories, such as 'sentence', 'noun phrase', 'adjective' and so on.

The rules of a generative grammar are rewrite rules, like those of [STR] and [BIN]. In the system [STR], the rewrite rules were *context dependent* – whether or not we could apply a rule to a symbol depended on surrounding symbols in the state. The system [BIN], however, had *context-free* rewrite rules.

A generative grammar has solely context-free rewrite rules which are such that there is only one symbol on the input side of any rule. A formal system which meets these criteria is called a *phrase structure grammar*.

Given a phrase structure grammar, we can generate all and only the grammatical strings according to that grammar by constructing *phrase structure trees*. A phrase structure tree is just like the generation trees we have seen so far with one exception. Where in the past nodes have contained states, the nodes of a phrase structure tree each represent only a single symbol, with its descendant nodes representing the symbol(s) with which it is rewritten. The grammatical strings given by a grammar are read off across the terminal nodes of a phrase structure tree.

Let's construct an example phrase structure grammar to make all this clearer.

14.3 PHRASE STRUCTURE TREES

We're going to specify a phrase structure grammar for a fragment of English. The states of our phrase structure grammar will be finite strings of those symbols which feature in the rules. The initial state will be the symbol 'S'. The rules of the system are as follows.

S	$\rightarrow$	S Con S / NP IVP / NP TVP NP
Con	$\rightarrow$	and / or / but
NP	$\rightarrow$	Det N
Det	$\rightarrow$	the / a
N	$\rightarrow$	Adj N
N	$\rightarrow$	man / woman / kitten / dog
Adj	$\rightarrow$	Adj Adj
Adj	$\rightarrow$	young / happy / cute / silly
IVP	$\rightarrow$	IVP Adv
IVP	$\rightarrow$	runs / eats / plays / smiles
Adv	$\rightarrow$	quickly / nicely / happily
TVP	$\rightarrow$	loves / disgusts / wants
PP	$\rightarrow$	P NP
P	$\rightarrow$	to

We can use this phrase structure grammar to generate grammatical strings, as represented by the terminal nodes of the phrase structure trees shown in Figures 14.1, 14.2 and 14.3.

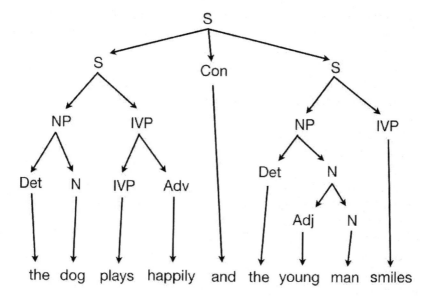

Figure 14.1   Phrase structure tree.

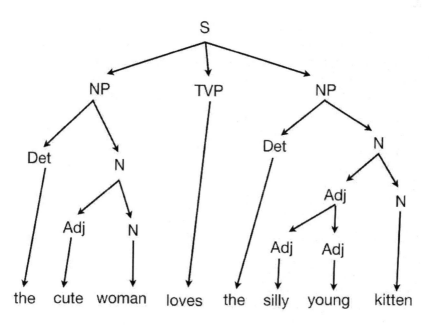

Figure 14.2   Phrase structure tree.

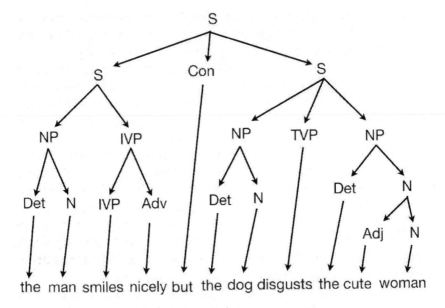

Figure 14.3   Phrase structure tree.

14.4 COMPUTING LANGUAGE

Determining the grammaticality of sentences of language according to a generative grammar is clearly a computational procedure. It should also be clear that given a phrase structure grammar and an arbitrary string of its symbols, we can conduct a bottom-up search to determine whether or not the string is generated by a phrase structure tree.

The example phrase structure grammar we constructed in the previous section is, of course, greatly simplified and considers only a small fragment of the lexical items and syntactic structures of English. Constructing a full generative grammar for a natural language involves not just specifying the rules by which phrase structure trees are constructed, but also specifying the various syntactic transformations on terminal strings of phrase structure trees which account for the myriad sentences native speakers produce.

This further element of syntactic transformation need not concern us here. As always, the interested reader can follow the suggestions for further reading or take an introductory course in generative grammar. It suffices for our purposes to make the following observations.

It seems that the mechanisms which facilitate grammaticality judgements in native speakers are computationally implementable. Given that this is one of the functions implicated in the various representational transformations involved in comprehending an utterance of natural language, we have made some small progress towards a computational account of the linguistic facility.

It should be clear to you now, however, just how complicated linguistic behaviour is to account for. Although producing and comprehending written and spoken language is so natural to us as to appear to be the most simple of processes, there are, in fact, a large number of mediating procedures facilitating linguistic activity.

While we have seen – at least in part – how one of these procedures might be accounted for in computational terms, there are still numerous mechanisms we are in want of an account for. We will return to some of these in later chapters. In Chapter 16 we are going to draw out further evidence that linguistic behaviour is rule governed and, hence, computationally implementable. In Chapter 19 we are going to examine how we might model some of the mechanisms implicated in reading written language.

For the moment, however, let's finish this chapter with one final problem. It appears that one of the most difficult elements of language comprehension to account for is the determination of semantics – *meaning* – from syntactic structures. Consider, for instance, the fact that many English sentences are *amphibolous*.

Amphiboly is a property of sentences such that they admit of more than one semantic interpretation as a result of their syntactic structure. This is distinct from cases of *lexical ambiguity* where a homonym introduces the potential for multiple interpretations. Sentence [8] below is lexically ambiguous, whereas sentence [9] is amphibolous:

[8]   The bank provided the pilot with a challenge.
[9]   I saw the man on the hill with the telescope.

While sentence [8] can be interpreted three ways, depending on the semantic interpretation of 'bank', sentence [9] admits of multiple interpretations but this ambiguity is not parasitic on the ambiguity of any particular word in the sentence. Rather, there are a number of ways in which we can interpret the syntactic structure of the sentence and these give rise to distinct meanings.

This problem of the determination of semantics from syntax is one we will return to at length in Chapter 17.

CHAPTER 15

HUMAN REASONING

In the previous two chapters, we approached the rational and linguistic faculties with a view to analysing their constituent mechanisms and accounting for these mechanisms in computational terms.

In this chapter and the next, we are going to examine evidence from psychology and linguistics that bears on the question of whether human rationality and linguistic competence are effectively rule governed and, hence, computationally implementable.

You will be aided in our brief examination of empirical data concerning human rationality in this chapter if you first answer the following reasoning problems:

1. If Mike is married then he is happy. Mike is married. Does it follow that he is happy?
2. No hippies are financial advisors. No financial advisors are nuclear protestors. Does it follow that some hippies are nuclear protestors?
3. There are four cards in front of you. Each card has a letter on one side and a number on the other. The cards are lying with one face up such that you can see the following on the showing faces:

   A K 4 7

   You are told that this rule holds of these cards – *If there is an A on one side, there is an even number on the other side*. Which of the four cards *must* you turn over to determine whether or not this rule does hold?
4. Adrian has wealthy parents and went to an elite grammar school. He works as an investment banker, is married to a corporate executive and drives a Porsche. Which of the following is more likely?

   (a) Adrian is concerned about public health and welfare.
   (b) Adrian is concerned about public health and welfare but votes conservative.

5. All the people in Queensland wear hats in the sun. No one who wears a hat in the sun gets facial melanomas. Does it follow that no one in Queensland gets facial melanomas?

6. A flipped coin has landed on heads nine times in a row. What are the odds it will land on tails on the next flip?

   (a) Even
   (b) Better than even
   (c) Less than even

7. Jon and Nicole have two children. One of them is a boy. What is the chance that the other one is a boy?

   (a) 50%
   (b) 33.33%
   (c) 25%

8. All the vegetarians on campus are members of the Organic Food Cooperative. None of the vegetarians on campus purchase food known to be treated with pesticides or preservatives. Does it follow that none of the members of the Organic Food Cooperative purchase food known to be treated with pesticides or preservatives?

9. You draw a card from a standard deck. Which of the following is more likely?

   (a) You draw an ace.
   (b) You draw a red ace.

10. You are given eight cards from a standard deck and are told that *only* one of the following two statements is true:

    [1] There is either an ace or a king in your hand (or both).
    [2] There is either an ace or a queen in your hand (or both).

    Which of the following is more likely?

    (a) There is a king in your hand.
    (b) There is an ace in your hand.

11. If you drink coffee at lunchtime then you will be more alert in the afternoon. Dave is not more alert this afternoon. Does it follow that Dave didn't drink coffee at lunchtime?

12. The rule for collecting unemployment benefits states – *If you collect benefits then you must not be employed.* Which of the

following four people *must* you gather further information about
to make sure they are not breaking the rule?

(a) The person you know to be employed.
(b) The person you know to be unemployed.
(c) The person you know to be receiving benefits.
(d) The person you know to be not receiving benefits.

15.1 FOLLOWING LOGICALLY

It is a recognised empirical fact that people generally perform quite
poorly on this set of reasoning problems. The reader of this volume
is clearly more intelligent than average – simply by virtue of having
purchased this book if nothing else – but I would still be surprised if
you made no mistakes on the problem set (unless you've previously
been exposed to these problems).

The question of whether a conclusion *follows logically* from some
premises is spelled out in terms of the *validity* of the inference. An
inference is valid iff the truth of its premises is sufficient to guarantee
the truth of its conclusion. The validity of a particular inference
depends on the validity of the logical form which it instances.
Technically speaking, it is logical forms of inference which are valid
or not. A logical form is valid iff there is no instance of the form which
has true premises and a false conclusion. In other words, a logical
form is valid if there is no instance of the form which is a *counter-
example* to its validity.

I still don't expect you to have a good understanding of the concept
of logical form or what it is precisely to instance a logical form, but it
is not necessary for present purposes that you do. I merely want to
make the point that whether or not a conclusion follows from some
premises – whether or not the inference is valid – is a purely formal
consideration. Determinations of the validity of inferences have
nothing at all to do with the actual content, or *meaning*, expressed by
the premises and conclusion.

It is demonstrably the case, however, that people generally *are* guided
by the meaning of premises and conclusions when making untutored
determinations of the validity of inferences. This is precisely why
experimental subjects make predictable errors in the problems above.

The primary aim of this chapter is to examine the kinds of errors
people generally make on such problems and to consider whether the
rational performance of logically untutored subjects on logical problems poses a challenge to computationalism.

15.2 RATIONAL PERFORMANCE

People generally spot the validity of simple inferences, such as in problem 1. This is just *modus ponens* so it does follow that Mike is happy. Problem 11 also instances a very simple inference form – *modus tollens*. Given a true conditional with a false consequent, we can always validly infer the falsity of the antecedent. However, it is common to see mistakes on this problem.

There are two reasons why this might be the case. One is that *modus tollens* involves negation and it seems that reasoning which includes negation is generally more difficult than reasoning which only involves affirming. Another reason is that, rather than assuming the truth of the premises in determining the validity of the inference, reasoning subjects are likely to think that there may be some complicating factor involved in Dave's afternoon sleepiness – perhaps he had a late night – and to thereby determine that the conclusion doesn't follow. In other words, they're likely to be guided by the meaning of the premises and the conclusion rather than their logical form.

This latter consideration doesn't actually speak against the validity of the inference though, merely against the truth of one of the premises. Whether or not the premises are actually true has no bearing on the validity of the inference – an inference is valid if the truth of the premises is sufficient to guarantee the truth of the conclusion. Valid inferences can have false premises and problem 11 is just such a case – lunchtime coffee consumption is no guarantee, in and of itself, of afternoon alertness, so the conditional is actually false.

Problem 10 gives further evidence that negation complicates reasoning tasks. The most common answer to this problem is (b) but the answer is, in fact, (a). Given the problem information, it is not possible for there to be an ace in the hand.

You are told that *only one* of [1] and [2] is true, which means that one of the statements is false. If it is [1] that is false then there is neither an ace nor a king in your hand. If it is [2] that is false then there is neither an ace nor a queen in your hand. So whichever statement turns out to be true, there is not an ace in your hand.

One way of accounting for the typical mistake is to note that people generally disregard the negative information and concentrate on the positive. So rather than thinking about what the falsity of one of the statements would entail, they concentrate on what the truth of either statement would entail.

The other part of the explanation for the typical mistake is that there being an ace in the hand is something which features in both statements, whereas there being a king in the hand features in only one. So if we are concentrating on what the truth of one of the statements would entail, we are likely to think that the truth of either would mean there might be an ace in the hand, whereas only the truth of one would mean there might be a king in the hand, and to thereby reason – erroneously – that it is more likely that there is an ace in the hand.

People also generally perform quite poorly on categorial reasoning tasks, as problems 2, 5 and 8 demonstrate. Of these, problem 5 is the simplest and the most likely to be correctly answered. It does follow from the premises that no one in Queensland gets facial melanomas but we might be led astray even in this simple case if we have background knowledge of the incidence of skin cancer in places that are subjected to harsh sun conditions.

Problem 2 is likely to be answered in the affirmative, but the correct response is that the conclusion doesn't follow in this case. The premises tell us nothing either way about the relation between hippies and nuclear protestors so, while it doesn't follow that some of the former are the latter, neither is it ruled out. Part of the explanation for the typical error here appeals to difficulty in reasoning about negative categorial relations, but an important part of the explanation is that the stereotypical hippy *would* be a nuclear protestor and this background information is brought to bear if we consider the meaning of the premises and the conclusion rather than just their logical form.

Problem 8 is also likely to be answered in the affirmative although the correct response is that the conclusion doesn't follow. The reasons for this are precisely those which account for the typical error in problem 5. It has partly to do with the fact that the reasoning involves negative categorial relations but is largely to do with the fact that the stereotypical Organic Food Cooperative member *would* avoid such foodstuffs.

Reasoning about probability also leads to characteristic errors, as problems 6 and 7 demonstrate. Those who answer 'better than even' to problem 6 fall prey to the *gambler's fallacy*. The gambler's fallacy is the view that the odds 'even out' over any course of trials. This, of course, is false. The odds of a fair coin landing on heads are always even, despite the results of any number of preceding trials. Flipping ten heads in a row is no more nor less likely than flipping any other combination of heads and tails.

The answer to problem 7 is, somewhat counterintuitively, that there is a one in three (33.33 per cent) chance that their other child is a boy.

Pretty much everyone gets this one wrong. The reason the answer is not 50 per cent is that we don't know *which* of their children is a boy and this affects the probability space.

There are four ways that Jon and Nicole could have two children. They could have a girl and then a boy, a girl and then another girl, a boy and then a girl, or a boy and then another boy. If all we know is that *one* of their children is a boy – and we don't know which one – then the only situation that is ruled out is the one in which they have two girls. This leaves three possibilities, one of which is such that their other child is also a boy, so the probability of this being the case is one in three.

If we knew that their *first* child was a boy, or that their *second* child was a boy, this information would rule out two of the possibilities, making the probability of the other child being a boy one in two. As it is, we only have enough information to rule out one of the four possibilities.

Problems 4 and 9 are of particular interest as both problems are structurally identical but untutored solutions to the problems typically diverge. No one ever answers (b) to problem 9 but people often answer (b) to problem 4. In both cases, (b) is the incorrect answer since a conjunction is never more probable than either of its conjuncts. While people are quick to recognise that drawing an ace is more probable than drawing a red ace, they are generally led astray by their background knowledge with respect to the information described in problem 4.

It seems as if the reasoning process people engage in with respect to problem 4 involves, once again, appeals to stereotypes. A stereotypical investment banker with elite grammar schooling and a Porsche in the garage is not the kind of person we expect to be concerned about public health and welfare. If, however, they tend to vote conservative despite these concerns, this makes for a closer – although still somewhat anomalous – fit to the stereotype. As such, we are inclined to think that (b) is the more likely case, given what we're told about Adrian.

Problems 3 and 12 are also especially interesting as they too are structurally identical – they have precisely the same logical form – yet they are typically differently answered. The most common answer to problem 3 is that we need to turn over the card showing 'A' and the card showing '4'. The correct answer, however, is that we need to turn over the card showing 'A' and the card showing '7'.

The reason is that in order to determine whether or not a rule holds, we need to look for disconfirming instances, not confirming instances. In the absence of counter-examples we can say that the rule holds – if we find a counter-example we have proof that it does not.

In the problem case, we need not turn over the card showing '4'. If there is an ace on the other side then it merely confirms the rule but if there is not, it does not provide a counter-example. The rule says only that if there is an A on one side then there is an even number on the other side. It doesn't say what must be the case if there is an even number on one side.

We do, however, need to turn over the card showing 'A' to make sure that there is an even number on the other side. We also need to turn over the card showing '7' in order to make sure that there is *not* an 'A' on the other side, as this would be a counter-example to the rule and would thereby show that it does not hold.

While people almost always get problem 3 wrong, they almost always get problem 12 right, yet this is precisely the same problem. The correct answer to problem 12 is that we need to check the person who is employed – to make sure they are not also collecting benefits – and the person who is collecting benefits – to make sure they are not also employed.

We can account for this performative contrast on structurally identical reasoning tasks by appealing again to the reasoning subject's background knowledge with respect to the problem information. Most of us know precisely what a welfare cheat is and implicitly understand that it is someone who breaks the rules. Consequently, we know exactly what to look for in question 12 – possible cases of rule breaking.

Cards with numbers on one side and letters on the other side are, in contrast, not something that most of us would ever have run across. As such, there is no relevant background information to tell us that we should be looking for cases of rule breaking to solve problem 3.

15.3 MENTAL MODELS

It seems we have plenty of evidence that when people reason, they do not ordinarily explicitly follow formal rules. Rather, they construct mental models of the problem situation and interrogate these mental models to determine a solution.

These mental models are sometimes constructed entirely – and selectively – on the basis of information given in the problem but often also appeal to relevant background information and comparisons to stereotypical or *paradigm* cases.

In answering problem 2, for instance, the mental model we construct involves a hippy who is not an investment banker and we tend to interpret the question of whether *some* such hippies are nuclear protestors as the question of whether it is probable or possible that the paradigm such hippy would be a nuclear protestor. Consequently we erroneously answer in the affirmative.

Similarly, in answering question 4, we construct a mental model of an investment banker with a privileged background and affluent lifestyle. We then weigh the paradigm such person against the likelihood of their social concern and find that while such social conscience does not fit with the stereotype, the addition of a conservative voting preference makes for a somewhat closer fit. Consequently we erroneously think that the model which also includes a conservative voting preference is the more likely one.

Although the construction of these mental models can sometimes lead us astray, they can also sometimes lead us very quickly to the correct answer, as in problem 12. From an evolutionary perspective, it is to be expected that humans would develop reasoning procedures that require as little cognitive resources as necessary and which place importance on past experience of similar situations. So it is to be expected that people often disregard certain problem information in order to simplify wherever possible and that they appeal to seemingly relevant background information, even though only the formal properties of the problem are strictly relevant.

If this means that we sometimes perform poorly on fairly artificial formal tasks, this is a small evolutionary price to pay for quickly and cheaply (cognitively speaking) getting it right most of the time in real-world reasoning tasks.

This claim that typical human reasoning involves the construction and interrogation of mental models is one theory which coheres and explains typical performance on these reasoning tasks. Although it is only a theory, it is one which is quite intuitive and which enjoys some currency, in one form or another and under various names, in cognitive psychology. The progenitors of this kind of theory, as it applies to these reasoning tasks, were Johnson-Laird, Tversky and Kahneman.

15.4 EXPLANATORY BURDEN

If it were the case that people always reasoned formally, according to the dictates of some logic, then providing a computational account of rational mechanisms would be very straightforward, since logics just are formal systems.

In light of this empirical data concerning typical performance on reasoning tasks, however, the computationalist is faced with an explanatory challenge – to account for the construction and interrogation of mental models in computational terms.

There is no prima facie reason, however, nor any reason we can derive from the empirical data, to suppose that these reasoning mechanisms cannot be accounted for computationally. If we weren't being philosophically careful, we might be tempted to license an argument against computationalism such as the following.

The reasoning processes of untutored subjects demonstrably do not simply involve explicitly following logical rules

---

∴ People typically reason illogically, or irrationally

∴ These reasoning mechanisms cannot be accounted for in terms of formal systems

---

∴ There is at least one mental process which is not computationally implementable

---

∴ Computationalism is false

The above argument begins to go wrong at the very first inference. There are strong and weak senses of 'irrational' that we must be careful not to equivocate on. We charge someone with being 'irrational' in the weak sense if their reasoning is guided by a principle which is, in fact, false. Someone who falls prey to the gambler's fallacy is a paradigm case of someone reasoning irrationally in the weak sense.

This weak sense of 'irrational', however, is not sufficiently strong to warrant the next inference in the argument against computationalism. The sense of 'irrational' that is imputed to untutored subjects such that we would concede the next step in the above argument is a much stronger sense. We would charge someone with being 'irrational' in this stronger sense if their reasoning was not guided by any principle at all, or if, in the face of an argument such that they endorse the truth of all the premises and the validity of the reasoning, they still refuse to accept the conclusion.

It is not the case, however, that the reasoning performance of typical subjects is such as to warrant the indictment of this strong sense of 'irrational'. In all cases, there is good evidence to suppose that the reasoning processes deployed *are*, in fact, guided by certain principles – they are just not formal principles which are guaranteed to be truth preserving.

Furthermore, when the correct methods of reasoning are explained to subjects who made incorrect determinations on reasoning problems, they are generally quite quick to spot their mistake and are not likely to make the same mistake again in future tasks. They might be initially resistant to accepting the correct conclusion – particularly with problems 3 and 7 – however, once the reasoning is properly explained (perhaps with diagrams or by reference to analogous situations) this resistance is overcome.

Consequently, the empirical data is not sufficient to warrant the strong claim that human reasoning mechanisms *cannot* be accounted for in terms of formal systems – a claim whose truth would demonstrate the falsity of computationalism.

Certainly human reasoning mechanisms cannot be accounted for purely in terms of *explicitly following* formal rules, but this is not *ipso facto* proof that the mechanisms involved are not implicitly governed by computational methods. Analogously, judgements of the grammaticality of sentences do not involve *explicitly* following formal rules, but this does not show that such judgements are not underwritten by computational processes.

So the challenge posed to computationalism by this empirical data is not insuperable and is restricted to the explanatory burden of giving a computational account of the mechanisms involved in typical reasoning.

A central mechanism that seems to be implicated in such reasoning is the ability to make comparisons to past situations and to paradigm cases. This involves recognising that the problem situation involves a known pattern of experience and invoking that known pattern to determine information relevant to the task.

This pattern matching and reconstruction is a mechanism that we may be able to account for very well with the systems we will examine in Chapter 19.

CHAPTER 16

HUMAN LANGUAGE

The human linguistic capacity is really quite amazing. The mechanisms which facilitate linguistic production and comprehension are surprisingly complex given that our capacity for language is so natural to us as to appear incredibly simple. No doubt you've begun to appreciate just how much cognitive processing is involved in linguistic behaviour after reading Chapter 14.

As difficult and complex as it is to theorise about the linguistic capacity, a child of six has already fully internalised the syntax, morphology and phonology of their first language. This has led linguistic researchers, following Chomsky, to postulate the necessity of some innate mechanism which aids in the acquisition of our first language.

There is much to be said about this postulated innate mechanism and its role in subserving the acquisition of language, but there is very little in this debate that bears on the tenability of computationalism, so this is not what I want to focus on in this chapter.

The aim of this chapter is to draw out evidence that much of our linguistic activity is strictly rule governed – hence computationally implementable – despite our ignorance of these rules. This will hopefully also lend weight to the claim of the previous chapter that even cognitive mechanisms which do not involve explicitly following rules can be accounted for computationally.

There is considerable evidence in favour of the view that linguistic behaviour is entirely rule governed. Much research in linguistics involves making explicit and codifying these implicit rules.

One area of linguistic study in which the identification of the implicit rules governing our behaviour is strikingly demonstrable is the study of *phonology*.

16.1 OBSTRUENT PHONEMES

The study of phonology is the study of the speech sounds and sound patterns of spoken language. Central to the study of phonology is the identification and classification of the phonemes of a given language.

We learned a little bit about phonemes in Chapter 14 where we saw that phonemes are the smallest units of speech that provide distinctive contrast. We're now going to build on this understanding and taxonomise the phonemes of English.

Phonemes divide into open, or *sonorant*, sounds – which we can think of as vowels – and restricted, or *obstruent*, sounds – which we can think of as consonants.

Obstruent phonemes are described in terms of their place and manner of articulation, and whether or not they are voiced. The place of articulation refers to the combination of articulatory apparatus that is employed in their production. The manner of articulation refers to the extent to which the sound is restricted by the articulatory apparatus.

The chart in Figure 16.1 taxonomises obstruent phonemes according to their voicing and manner of articulation along the vertical axis, and their place of articulation along the horizontal axis.

Before reading on, consult the pronunciation chart provided in Figure 16.2 and practise producing each of the phonemes, concentrating on the placement of your lips, teeth and tongue and the extent to which the passage of air is restricted by this placement in producing each phoneme.

	Bilabial	Labiodental	Interdental	Alveolar	Alveopalatal	Velar	Glottal
Voiced stops	b			d	g		
Voiceless stops	p			t	k		
Nasals	m			n		ŋ	
Voiced fricatives		v	ð	z	ʒ		
Voiceless fricatives		f	θ	s	ʃ		h
Voiced affricates					ʤ		
Voiceless affricates					ʧ		
Approximants	w			r l		j	

Figure 16.1 Taxonomy of English obstruent phonemes.

b	bat	ʒ	asian
d	dot	f	fit
g	got	θ	thin
p	pot	s	sat
t	tot	ʃ	shy
k	cat	h	hot
m	mat	dʒ	jury
n	net	ʧ	chat
ŋ	sing	w	wet
v	vet	r	rat
ð	that	l	lot
z	size	j	yacht

Figure 16.2   Pronunciation chart for English obstruent phonemes.

Stops, or plosives, are phonemes such that their production involves completely obstructing the passage of air through the articulatory apparatus. Plosives can be voiced, such as /d/, or voiceless, such as /t/.

Fricatives are phonemes whose production involves restricting the passage of air through the articulatory apparatus such that the passage of air creates a sibilant, hissing sound. Fricatives can also be voiced, such as /z/ or voiceless, such as /s/.

Nasal sounds are those whose production involves the resonation of air in the nasal cavity. It is these obstruent sounds whose production is inhibited when we have a cold. If you hold your nose while trying to produce these phonemes, you will find that the sound is flat and wooden, whereas holding your nose has no effect on the production of any other obstruent phonemes. All nasal phonemes are voiced.

Affricate phonemes are combinations of plosives and fricatives. The passage of air is initially completely restricted, then partially restricted. There are only two affricates in English – the voiceless affricate that is the initial sound in the word 'chat' and the voiced affricate that is the final sound in the word 'fridge'.

Finally, we have approximants. The classification of these phonemes is somewhat contentious but the one above serves our purposes. All approximants are voiced. The approximant phonemes are the closest obstruents to sonorant phonemes.

This exhausts the manner of articulation of obstruent phonemes.

The other axis of classification is the place of articulation, which describes the primary articulatory apparatus involved in their production.

Bilabial phonemes are those which involve the use of both lips in their production. Labiodental phonemes involve placing the upper teeth against the bottom lip. Interdental phonemes involve placing the tongue between the teeth.

The alveolar ridge is the hard raised ridge that is just behind the upper teeth. Alveolar phonemes are those whose production involves placing the tongue against or near the alveolar ridge.

Continuing behind the alveolar ridge along the upper jaw is the hard palate. Alveopalatal phonemes are produced by placing the tongue behind the alveolar ridge and against, or close to, the hard palate.

If you run a finger along the roof of your mouth towards the back of your throat to the point where the hard palate ends, you'll feel the squishy soft palate. The soft palate is also known as the velum. Velar phonemes are those which are produced by placing the tongue near, or against, the velum.

The glottis is the space between the vocal chords. The glottal fricative /h/ is produced by restricting the passage of air through the glottis, but without vibrating the vocal cords.

Speakers of some dialects of English – such as the London cockney dialect – produce a glottal stop rather than the alveolar stop in certain phonemic contexts (such as in 'bottle' or 'sorted'). This is not a distinct English phoneme but an *allophonic variant* of the phoneme /t/. We will learn more about allophones shortly, but first let's taxonomise the sonorant (vowel) phonemes.

16.2 SONORANT PHONEMES

All sonorant phonemes are *ipso facto* voiced. They also all have the same manner of articulation – they are open sounds. In other words, the passage of air through the articulatory apparatus is not impeded but resonates freely in the oral cavity.

Sonorant phonemes are taxonomised along two axes which describe the position of the tongue in the mouth during their production. One axis corresponds with the height of the tongue and the other corresponds with the part of the tongue that is raised or lowered.

Unlike the obstruent phonemes which admit of clear discontinuities between phonemes given the distinct articulatory apparatus involved in their production, the sonorant phonemes are placed on a

Space of English Vowel Phonemes

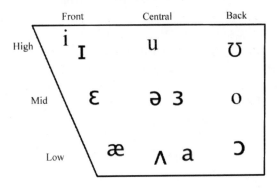

Figure 16.3   Vowel space.

continuum, since that the possible positions of the tongue in the mouth are continuous, not discrete. As such, vowel phonemes fall within a two-dimensional space of possible vowel sounds.

Consequently, we identify *cardinal* vowels within this vowel space as shown in Figure 16.3. The vowel space is continuous from the highest, most fronted sonorant phoneme, through to the lowest and least fronted sonorant phoneme.

The cardinal vowels represented in Figure 16.3 are all *monophthongs*. This means that their production involves a single continuous tongue position. As well as the monophthong phonemes, however, there are also *diphthong* phonemes. Diphthongs are phonemes whose production begins at one of the cardinal vowel positions but moves, during the production of the sound, towards the tongue position of another of the cardinal vowels.

The pronunciation chart provided in Figure 16.4 represents the vowel sounds in the dialect of Australian English that I speak. Note, however, that vowel pronunciation will vary between dialects of English, particularly with respect to the production of diphthongs, but also with respect to some monophthongs. Consequently, some of the triples in Figure 16.4 which all rhyme in my dialect of English may not rhyme in yours, depending on where you learned to speak English. Also, speakers of dialects of English other than Australian – particularly American dialects – are likely to produce monophthongs in place of some of the diphthongs.

Monophthongs

i	pea	feet	ease
I	pit	fit	is
ɛ	pet	fret	met
æ	pat	fat	mat
u	put	foot	should
ʊ	lute	fruit	shoot
ə	penc<u>i</u>l	the	eff<u>or</u>t
ɜ	purred	heard	shirt
o	pot	hot	shot
ʌ	putt	butter	mother
a	part	barter	father
ɔ	port	bought	fraught

Diphthongs

aɪ	I	my	sigh
ɛɪ	ate	weight	shave
ɔɪ	toy	oil	buoy
ɪə	here	hear	seer
ɛə	there	share	bear
ʊə	lure	tour	fewer
aʊ	loud	now	pout
oʊ	boat	stow	hoe

Figure 16.4 Pronunciation chart for Australian English vowel phonemes.

16.3 ALLOPHONES AND PHONETIC REALISATION

Phonemes are idealisations. Actual speech sounds – *phones* – approximate to phonemes and may vary significantly between speakers. As language users, we are very good at detecting the distinctive contrasts in speech sounds and assimilating phones to phonemes.

The first point of interest here is that typical monolingual speakers can only assimilate phones to the phonemes that are in their native language. This means that they literally can't hear phonemes that aren't in their language as they are not sensitive to phonetic contrasts that are not distinctive in their native language.

Since natural languages differ in the phonetic contrasts that are semantically distinctive, this can make learning a second language particularly difficult if the second language contains phonemes that are not in the first language. In Vietnamese and Mandarin, for instance, the tone of a speech sound is distinctive. Tone is not a distinctive contrast in English, however, so English native speakers learning these languages have enormous difficulty hearing the distinction as they automatically assimilate the relevant phones to the same English phoneme.

This is complicated by the fact that one and the same phoneme may be realised in distinct (but not semantically distinctive) *allophones* depending on the phonemic context of utterance. The phonetic difference between allophones is very difficult for a native speaker of the language to hear as the contrast is not distinctive but predictable from the context of occurrence.

Allophones occur in complementary distribution. This means that if a phoneme has more than one associated allophone, each allophone will always and only be produced in predictable phonemic contexts according to *phonetic realisation rules*. The phonetic realisation of allophones in complementary distribution is *strictly rule governed*. Some examples will serve to make this strikingly clear.

One phonetic realisation rule in English involves *aspirating* voiceless stops when they are word initial. This means that if a word begins with a voiceless stop, then the stop is produced with a little puff of air that is *not* produced in other contexts.

Put the palm of your hand right in front of your mouth and utter the words 'top' and 'stop' repeatedly and you should notice the difference. When you produce the word initial /t/ in 'top' you should feel the extra puff of air on the palm of your hand. When you produce the /t/ in 'stop', however, this puff of air is absent since this allophone of /t/ is not aspirated. Try this out with other pairs of words containing aspirated and unaspirated voiceless stop allophones, such as the /k/ in 'Cate' and 'skate' or the /p/ in 'pat' and 'apple'.

It is very difficult to actually hear the difference between these phones, however, precisely because they are allophonic variants – aspiration is not a distinctive contrast in English.

Another phonetic realisation rule of English is that vowels are nasalised always and only before nasal obstruents. Hold your nose while pronouncing 'cat' and 'can' repeatedly and you should be able to tell the difference. Again, try this out with other pairs of words containing nasalised and non-nasalised vowels, such as 'pot' and 'pond' or 'sit' and 'sing'.

Some phonetic realisation rules are restricted to certain dialects, like that which governs the production of the glottal stop allophone of /t/ in cockney English. Speakers of cockney English still produce the standard voiceless alveolar stop allophone of /t/ when it is word initial or follows a fricative, such as in 'tell', 'start', 'faster' or 'softer', but produce the glottal stop when /t/ occurs in intervocalic contexts (between vowels) such as in 'butter' or 'letter', and also when /t/ occurs word finally, such as in 'short' or 'let'.

The interesting point for our purposes is that this strictly rule-governed activity occurs despite our total ignorance of these phonetic realisation rules and our complete insensitivity to the phonetic distinctions between the allophones produced.

In fact, these phonetic realisation rules are so deeply ingrained that we can't help but carry them over illicitly when we learn a second language. We're typically unable to notice that we're actually producing different sounds in different contexts but it is likely to be apparent to a native speaker of the language we are learning.

An example of this is that English native speakers learning French will continue to (incorrectly) aspirate word initial voiceless stops – such as the /t/ in 'Tour de France' – even after this has been pointed out to them. As speakers of a language where this aspirated phoneme is in complementary distribution with other allophonic variants, we are simply unable to detect that we're doing it. Similarly native speakers of German learning English have a tendency to devoice word final stops – producing the /g/ in 'dog' as /k/ for instance.

So, not only do we tend to try and produce the sounds of our second language using only the phonemes of our first language, we also carry over our native phonetic realisation rules into our second language. These two factors account for the accent that a non-native speaker will typically never quite be able to get rid of. They may eventually – depending on the first and second languages in question – be able to accurately produce the phoneme set of the second language, but overriding phonetic realisation rules is very difficult. This is also the case with dialectical variations of the same language, such as the many dialects of English.

Fortunately, processing accents is also something we are generally quite good at. Sometimes it may take some exposure before we are able to process accents with ease – I found the Glaswegian accent to be impenetrable at first – but with sufficient exposure, we implicitly learn to apply a filter to the sounds we are hearing which assimilates them to the intended phonemes. While we might have to explicitly focus on this initially, we quickly internalise these transformative principles.

This indicates that we continue to internalise rule-governed linguistic principles long past the time at which we acquire our native language, even if we never learn a second language. Whenever we are first exposed to speakers of dialects of English that are distinct from our own, or speakers of English as a second language, we very quickly internalise rules that allow us to process their accent with ease. Of course, as far as the speaker of another dialect of the same language is concerned, it is us who has the accent – I'm sure Glaswegians find my Australian accent as initially impenetrable as I found theirs.

16.4 FIRST-LANGUAGE ACQUISITION

We can also find plenty of evidence of rule-governed linguistic activity in the first-language acquisition literature.

When a child reaches twelve months or so of age, they enter the single word stage of language acquisition. They are able to point at objects and name them with single word utterances, and they are also able to indicate some of their desires with single word utterances – 'ball', 'doll', 'mummy'.

At this stage it is common to see both semantic *overextension* and semantic *underextension* of newly acquired terms. Overextension, as the name suggests, is when a term is applied to referents beyond its proper extension. When, for instance, a child first acquires the word 'ball' they may well then apply this term to other round objects, such as some fruit or the Moon. Similarly, if they have a dog and learn its name, they may well then apply this term to all dogs.

Semantic underextension, as you have no doubt guessed, is the opposite phenomenon. Underextension occurs when a child restricts the application of a newly acquired term to only certain of its proper referents. They might, for instance, only use 'toy' for a particular favourite toy. Underextension tends to be less common than overextension.

In both cases, the child fairly rapidly learns the correct scope of application of the new term in their lexicon. This is at least prima facie evidence that the child is internalising the rules that govern the correct application of the term, based on observable features of its referents.

Another typical feature of the single word stage is the systematic phonemic substitution of certain phonemes that the child is unable to produce. For instance, the alveopalatal voiceless fricative is quite a difficult phoneme to produce as it requires fairly dextrous tongue placement so it is not uncommon to hear children systematically replace it with either the voiceless alveolar fricative or the voiceless interdental fricative – so 'ship' is pronounced 'sip' or 'thip'.

As well as making these phonemic substitutions, the child is also likely to make systematic phonological simplifications. The initial syllable in 'sleep' is also phonetically difficult to produce, so the child is likely to produce 'seep' in its place. More phonologically complex words are even further simplified – often idiosyncratically. A common example is 'sketty' for 'spaghetti'.

The particularly interesting point here is not just that children make these systematic substitutions, but that they are also sensitive to the fact that they are doing so. They are typically perfectly able to hear the difference between an adult's utterances of 'seep' and 'sleep' and, depending on their age, may well realise they are being teased if you reproduce their phonetic output rather than the correct utterance. So the development of their ability to comprehend phonemes outpaces the development of their ability to produce them and they compensate for this by making systematic – rule-governed – simplifications and substitutions.

16.5 LANGUAGE AND RULES

While we've mostly concentrated in this chapter on phonological processes, other areas of linguistics are also rich with examples of rule-governed behaviour. If you have some exposure to linguistics, or are planning to take an introductory course, I'd urge you to reflect on your knowledge – or to approach the subject – with a particular view to looking not just for evidence of rule-governed behaviour, but also for processes that might be troublesome to account for computationally.

In the preceding sections, we concentrated on identifying evidence in favour of the computational implementability of the linguistic facility. In the following chapter we are going to return to philosophical material and problematise a key aspect – arguably the most crucial aspect – of the production and comprehension of language, namely the determination of *meaning*.

CHAPTER 17

MEANING

This chapter marks a return to philosophical material after six chapters of technical material.

We've seen how computers can be programmed to strategically play complex games and we compared this to our intuitive understanding of how humans play these games. We've looked at expert systems as an example of machine reasoning and we've considered typical human performance on certain reasoning tasks in the context of determining the scope of the explanatory burden for the computationalist in accounting for this typical performance.

We've also examined how we might employ formal systems and search procedures to facilitate one of the mechanisms implicated in the linguistic facility – ruling on the grammaticality of strings. In addition, We've drawn out evidence from linguistics – mostly pertaining to phonological processes – which supports the notion that language behaviour is rule governed and, hence, computationally implementable.

Next we're going to consider a thought experiment which targets computationalism and seeks to show that there is a crucial facet of mental life that the computationalist cannot account for – the fact that our mental states are *meaningful*.

17.1 THE CHINESE ROOM

It is a crucial and defining feature of our mental states that they have semantic content – that they are *meaningful* states. Any adequate theory of mind must be able to account for the semantic contents of mental states.

Computation is an entirely syntactic process. The operations of formal systems are syntactically specified symbol manipulations. We've seen the explanatory efficacy of formal systems in accounting for a number of cognitive mechanisms. The crucial question for present purposes is whether or not we can account for semantics in terms of syntactic operations.

The thought experiment I want to entertain here was originally described by John Searle and seeks to establish that syntax alone is not *sufficient* for semantics.

Imagine that you are asked to spend several hours carrying out a certain task for experimental purposes. You are introduced to an enormous room containing thousands of shelves of numbered books. There is a table in the centre of the room with one of the books upon it. You flick through the book and see that it contains nothing but rewrite rules for symbols that you've never seen before, with a numerical notation alongside each rewrite rule.

You are told that you will be left alone in the room, at which point a piece of paper with a string of symbols on it will be passed through a slot in the door. Your task is to find the rule in the book whose input side is exactly that string of symbols and to copy out the output string of symbols onto the other side of the piece of paper and pass it back through the slot. You are then to find the book on the shelves whose number corresponds with the numerical notation alongside the rule you just followed and replace the book on the table with this new book.

You are left alone in the room and things proceed exactly as described. A piece of paper is passed through the slot in the door, you trawl through the book on the table to find this string on the input side of a rule and copy out the output string of the rule, then you replace the book on the table with the book from the shelves whose number was given by the notation next to the rule. Another piece of paper with a new string of symbols is passed through the slot in the door and you repeat the procedure.

After doing this for several hours, you are told that the strings of symbols were actually sentences in Chinese script. The books in the room encode all the possible conversations you might have in Chinese in several hours. Each book represents a conversation state and provides reasonable responses for possible inputs.

It turns out that you have been having a conversation with a Chinese native speaker for several hours and, on the basis of merely following the rewrite rules encoded in the books, have passed a Chinese Turing test.

Clearly, however, you do not thereby *understand* Chinese. For one thing, the conversational replies you were making to the Chinese questions did not accord with your beliefs and desires, but with arbitrary responses encoded in the books. For instance, one of the questions might have been 'do you like strawberry ice cream?' and your scripted response was 'yes, it's delicious', despite the fact that you can't abide strawberry ice cream. Or perhaps one of the questions was 'are you getting hungry?' and your scripted response was 'no, I'm fine for the moment, thanks', despite the fact that you were ravenous and wondering when lunch was.

For another thing, your capacity to converse in Chinese does not extend beyond the Chinese room. If a Chinese native speaker were to pass you a written Chinese query once you have left the room, the strings of symbols would still be meaningless to you. It is only through recourse to the encoded conversation states in the books of the Chinese room that you are able to give the appearance of under-standing and to pass the relevant Turing test.

The situation described in the thought experiment is one in which the processes of a formal system – the rewriting of symbols accord-ing to formal rules – suffice to pass a Turing test. The intuition that is primed by the thought experiment, however, is that even though the *appearance* of understanding is evidenced by the system – sufficiently well to convince a human – there is something crucial lacking in the operations of the system. They don't, in and of themselves, *mean* anything.

In other words, the syntactic operations of the Chinese room, although they pass a Turing test, lack *semantics*.

17.2 SYNTAX AND SEMANTICS

On the strength of the Chinese Room thought experiment, we might be tempted to mount this argument against computationalism:

P1 Having semantics is a necessary condition for having a mind.
P2 The syntactic operations of formal systems are not sufficient for having semantics.

∴ The operations of formal systems are not sufficient for having a mind.

∴ Computationalism is false.

Premise 1 is not in dispute – it is clear that mental states have seman-tic content. Premise 2, however, is certainly arguable.

There are two ways we might interpret the second premise, and, consequently, the interim conclusion that follows from it. The weaker interpretation is the claim that is licensed by the thought experiment;

however, a much stronger interpretation is appealed to in deriving the claimed falsity of computationalism.

The weak interpretation of the second premise is that there is a formal system such that its operations are not sufficient for having semantics. The stronger interpretation is that there is *no* formal system whose operations are sufficient for having semantics.

The Chinese room thought experiment does not show that there can be *no* formal system whose operations are sufficient for generating semantics. Consequently, the argument above fails to show the falsity of computationalism. What the thought experiment does show, however, is something a little stronger than the weak interpretation of the second premise I've given above.

The weak interpretation of premise 2 is essentially trivial. It's a given that there are plenty of formal systems whose operations don't meet conditions for having a mind. What is interesting about the Chinese room thought experiment, however, is that it shows that there is a formal system whose operations alone *are sufficient for passing a Turing test*, yet, intuitively, the system lacks understanding entirely.

We might, then, interpret the Chinese room thought experiment as an indictment on the efficacy of the Turing test. After all, if something can pass the test despite a complete lack of understanding, it doesn't seem the test is at all a reliable indicator of the presence of a mind. Before we draw this conclusion, however, we should reflect on the system described in the thought experiment in light of what we know about formal systems and natural language processing.

The thought experiment describes a system which, while *logically* possible, is not *physically* possible. To implement this system, we would need to draw up the generation tree for all possible Chinese conversations that can be had in the course of several hours.

Given that a generation tree for possible conversation states would be considerably more complex than the generation tree for possible chess states, it should be clear that constructing a complete generation tree for even the first twenty possible conversational exchanges is simply not computationally tenable. It is safe to say that, regardless of future advances in practical computational power, no computer will ever be able to pass a Turing test by following the method which the Chinese room implements.

Ordinarily, arguing against the physical possibility of a thought experimental situation obtaining does no philosophical work since, generally, we are using thought experiments to test claims of *logical* relations.

For instance, physicalism holds that a complete physical description is *sufficient* as a complete description of the mind. This sufficiency is a logical claim. Consequently, while it is physically impossible that there could be a scientist such as Mary, the thought experiment described in Chapter 5 still speaks against this sufficiency – to the extent that your intuitions are primed by the thought experiment – since the situation described is logically possible.

If the claim concerning the Turing test was similarly a *logical* claim, then the Chinese room thought experiment would indeed speak against it. Recall, however, that the claim is not that passing a Turing test is *sufficient* for having a mind, but rather that, were something to pass a Turing test, we should be prepared to attribute mentality to it.

The Turing test is an *empirical* test. Consequently, when considering possible counter-examples to the efficacy of the test as a reliable indicator of the presence of a mind, we should restrict our consideration to *empirically* possible systems.

So the Chinese room thought experiment fails to straightforwardly show the falsity of computationalism and offers no indictment on the efficacy of the Turing test. It does, however, still show something rather important.

The thought experiment does, I think, show that no amount of syntactic operation in isolation from the external world is sufficient for generating semantics. I could be in the Chinese room performing this procedure for years – given enough books – and it seems, intuitively, that there is no way to begin to understand the meaning of the symbols I am processing. The reason is that my operations lack an appropriate connection to the external world.

To understand Chinese *just is* to understand how elements of the language – written or spoken – relate to things *outside* the system of language. Languages are systems which encode and communicate meanings. These meanings, however, are not generated by mechanisms and inputs entirely internal to the linguistic facility. Certainly linguistic mechanisms are implicated in the conferral of meaning to linguistic entities but *necessarily* implicated is an appropriate connection to the external world.

The lesson to draw from the Chinese room thought experiment is that *embodied experience* is necessary for the development of semantics. In order for our mental states to have meaning, we must have antecedent experience in the world, mediated by our sensory apparatus. In other words, semantics do not develop in isolation but, rather, this development is conditional on experience in relation to the empirical world.

This necessity of embodied experience for the development of semantics does not, in and of itself, speak against computationalism. It merely shapes the explanatory burden on the computationalist, requiring them to provide a computational account of the meaning conferring mechanisms. This will involve, *inter alia*, an account of the computational conversion of sense data to various kinds of *mental representations*, which are then involved in further computational processes – such as the comprehension of natural language utterances.

There is a distinction to be drawn here between the conditions for the development of semantic representations and the conditions under which tokens of these representations are held to be meaningful. In other words, we might concede the necessity of embodied experience for the development of semantic representations, but then consider a thought experiment where functional equivalents of the formal system(s) facilitating language production and comprehension in a fully developed native speaker are enacted in a way approximating the Chinese room example, hoping to further problematise the meaningfulness of the operations of such a system.

Recall, however, from Chapter 14 that although there is no reason to suppose that these linguistic processes are not computational, there is good evidence to support the claim that concerted appeals to various kinds of mental representations – including semantic representations – are a necessary feature of their operations.

In other words, it is not the case that the processes underwriting linguistic comprehension are isolated, modular processes, with meaning being assigned as the final stage of processing. Rather, these various processes occur in concert with appeals to semantic representations serving to constrain and inform phonological and syntactic processes. Consequently – although I've offered neither proof nor robust argument here – it seems likely that any empirically possible system sufficient for passing a Turing test will necessarily already contain meaningful semantic representations.

There is much more to be said about the conditions under which operations of a formal system can be held to be intrinsically meaningful (not just interpretable as meaningful). In the following chapter, I want to shift the focus of our examination of meaningfulness to explicit consideration of this notion of *mental representation*.

CHAPTER 18

REPRESENTATION

Our mental states are meaningful by virtue of being *about* things. In other words, meaningful mental states are *representational* states – they *represent* or *stand for* things. In previous chapters I've made reference to *mental representations*, such as the phonemic, syntactic and – crucially – semantic representations which facilitate linguistic production and comprehension. In this chapter, I want to briefly discuss the structure and nature of mental representation.

Representation is quite a thorny philosophical topic. I don't intend this chapter to be a comprehensive introduction to the various debates concerning mental representation. Rather, I want to use this discussion of the nature of mental representation to introduce a distinction between two competing paradigms in artificial intelligence research.

The most important distinction, for our purposes, between these two paradigms – which I will call the *symbolic* and the *connectionist* paradigms – lies in the methods deployed by researchers in trying to computationally replicate cognitive functions. The symbolic artificial intelligence researcher will employ symbol systems of the kind we are now very familiar with, having seen numerous examples in previous chapters. The connectionist artificial intelligence researcher, on the other hand, will construct *artificial neural networks*.

We're going to examine artificial neural networks – or *connectionist networks* – at length in Chapter 19. In what follows, I want to make clear that this distinction in artificial intelligence methodology is partly (although not entirely) informed by beliefs concerning the nature of mental representation.

18.1 INTENTIONALITY

Intentionality is the technical philosophical term for the representational nature of mental states. *Intentional* states are those which are *about* something, which *represent* something.

The terms 'intentionality' and 'intentional' are not to be confused with the verb 'intend' and its cognates. Whether or not a state is intentional, in the technical philosophical sense, has nothing to do with it being intended by some agent. Rather, a mental state is intentional just in case it is *about* something. Intentionality is the property of mental states such that they are directed towards an object of representation (a thing which is represented).

It is mental representations which are the primitive bearers of intentionality. Our mental states are intentional by virtue of having mental representations as constituents. So, for instance, my belief that 'my dog is a fine companion' involves, *inter alia*, a token mental representation of 'my dog'. My belief is *about* my dog by virtue of having a constituent mental representation which is about my dog. My mental representation of 'my dog' is directed towards (about) its object of representation – namely, my dog.

This brings us to one of the important features of mental representations I want to highlight here. Mental representations are *categorial*. My mental representation of 'dog' picks out all and only dogs. In other words, it serves to *categorise* – in my mental life – those things which I take to be dogs and distinguish them from those things which I take not to be dogs. Similarly, my mental representation of 'brown' picks out all and only the things I take to be brown and my mental representation of 'chair' picks out all and only the things I take to be chairs.

The other important feature of mental representations I want to highlight is that they are *compositional*. Mental representations *compose* into more complex mental representations. Given, for instance, my possession of mental representations of 'brown' and 'dog', I need nothing further to compose the more complex mental representation of 'brown dog'. This more complex mental representation picks out all and only the things I take to be brown dogs.

This *compositionality* of mental representations allows for one part of an account of how it is that mental representations are conferred with their intentional content. Complex mental representations inherit their intentionality from the primitive mental representations of which they are composed.

The crucial – and most difficult – question to answer, however, is how it is that *primitive* mental representations are conferred with their intentional content. How is it that our atomic mental representations come to be *about* their objects of representation? In other words, what is the nature of the relation between mental representations and the (categories of) objects they represent?

18.2 CATEGORIES AND CONTENT

To give an account of the relation between mental representations and their intentional objects is to give part of a semantics for mental representation. There are numerous theories of the semantics of mental representation but I'm not going to give a balanced exposition of the available theories here. Instead, I want to give just the barest sketch of two *kinds* of theories.

On one hand we have theories according to which mental representations are essentially discrete. On the other hand we have theories according to which mental representations are essentially interrelated.

Theories of this first kind fit well with the symbolic paradigm in artificial intelligence research. According to this kind of theory, mental representations are symbols.

Theories of the second kind fit well with the connectionist paradigm in artificial intelligence research. According to this kind of theory, mental representations are distributed patterns.

A commitment to symbolic representation, on the one hand, or distributed representation on the other, brings with it a raft of corollary commitments. These include commitments concerning the mechanisms by which representations are conferred with their intentional content, the nature and structure of the categories represented and the ways in which mental representations interact.

Proponents of symbolic representation take representations to be essentially discrete in a number of ways. The mechanism by which symbols are conferred with their content is understood as some kind of direct relation between tokens of the symbol and objects of representation. Crucially, this mechanism is such that the content of a symbol in no way depends on the content of other symbols. Each symbol is discretely conferred with its intentional content.

Furthermore, symbolic representations are understood to remain discrete in their interactions with other representations. The compositionality of mental representation is understood to be simple syntactic concatenation. When symbols compose to give more complex representations, each symbol always brings the same content to the complex in which it participates. In other words, the content of symbols is taken to be contextually insensitive.

It is a further feature of symbols that their presence is binary – a symbol token is either present or not. If a symbol token is present, it is fully present and if it is absent, it is fully absent. Symbols, if you will, are either on or off, with no scope for anything in between. This binary nature of symbolic representation has implications pertaining to the nature of the categories which they represent.

If mental representations are symbolic then the categories which they represent must admit of sharp borders and no internal structure. In other words, if mental representations are symbolic, then the categories they represent are like boxes – objects are either in the category or not and there are no better or worse cases of category membership.

Proponents of symbolic representation, to recap, take the content conferring mechanism on representations to be discrete, the categories they represent to be binary and unstructured, and the composition of mental representation to be contextually insensitive syntactic concatenation.

Advocates of distributed representation, on the other hand, have a different understanding of each of these elements of the semantics of mental representation.

18.3 SYMBOLS AND PATTERNS

Theorists who endorse an account of mental representations as distributed patterns understand representations to be essentially interrelated. The semantics of mental representation that such a theorist will advance are such that the mechanism by which content is conferred on a representation is essentially mediated by relations to other representations.

There are a number of ways to flesh out this mediation but the mechanics of particular theories need not concern us here. What is important for our purposes is the commitment to the interrelated nature of mental representations, in stark opposition to the view held by proponents of symbolic representation. The way in which a mental representation is conferred with its intentional content, according to distributed accounts of representation, is essentially bound up with the way in which other mental representations are conferred with their intentional content.

The composition of distributed mental representations is also taken to be somewhat more complex than syntactic concatenation. It is crucial that there be an account of the composition of mental representation that is sufficient to secure the systematicity of cognition, since this is generally held to account for the productivity of the linguistic facility.

Those who endorse a view of mental representation as distributed understand the composition of representations to be the highly complex interaction of patterns of activation in a distributed network. Precisely what this means will become clearer in the following chapter when we discuss artificial neural networks. For present purposes, it suffices to appreciate that the way distributed representations compose is contextually modulated. In other words, the content that a particular representation brings to the complex representation in which it participates will vary in a way that is dependent on the other particular representations also participating in the complex.

It is a further feature of distributed representations that they can be partially tokened. Since token representations are taken to be patterns of activation widely distributed across an interconnected network of nodes, these patterns can be partially activated. Again, precisely what this means will become clearer in the following chapter. The important point here is an appreciation of the implications that the possibility of partial tokening of mental representations has with respect to the nature of the categories represented.

If mental representations are distributed patterns which can be partially tokened, then the categories they represent can admit of imprecise borders and internal structure. Furthermore, if the content of representations is contextually modulated, then the extension of the category will be contextually sensitive.

In other words, if representations are distributed and contextually modulated, then the categories they represent are such that there can be borderline cases of membership, the borders can shift contextually and there can be graded membership admitting of better and worse cases.

To recap, advocates of distributed representation take the content conferring mechanism on representations to be essentially mediated by relations with other representations, the categories they represent to be contextually sensitive – allowing imprecise and shifting borders and internal structure – and the composition of mental representation to be the complex, contextually modulated interaction of patterns of activation in a highly interconnected network.

18.4 COGNITIVE ARCHITECTURE

So far in this chapter I've discussed two distinct views of mental representation and used this distinction as an entryway into understanding the competing symbolic and connectionist paradigms in artificial intelligence research.

These differing views concerning mental representation are of central importance in distinguishing between the two paradigms but they do not exhaust the differences between them. Connectionists also differ from their symbolic counterparts with respect to views concerning *cognitive architecture*.

The term *cognitive architecture* refers to the structure and nature of the information processing systems of a cognitive agent. In other words, the term refers to the organisational and implementational features of the computational hardware which facilitates cognition.

The symbolic tradition in artificial intelligence research sees the cognitive architecture of the human mind as a *physical symbol system*. Connectionists, on the other hand, view human cognitive architecture in terms of *connectionist networks* which facilitate *parallel distributed processing*.

In previous chapters we've seen numerous examples of how we might implement cognitive functions with symbol systems. Connectionist networks, as we will see in the following chapter, are particularly well suited to carrying out functions that are notoriously difficult to implement in symbol systems architecture.

To the extent that connectionist architecture is readily amenable to implementing functions which we take to be importantly constitutive of cognition and which prove problematic to implement with symbol systems, we have at least one reason for preferring a connectionist approach over a symbolic approach to artificial intelligence.

The following chapter will be devoted to making clear the concepts which have so far only been mentioned with little in the way of explanation. After explaining these concepts and exemplifying the operations of connectionist networks with numerous examples, we will then return to further discuss the relation between the symbolic and the connectionist paradigms.

CHAPTER 19

ARTIFICIAL NEURAL NETWORKS

The connectionist paradigm in artificial intelligence research rose to prominence in the last two decades of the twentieth century. Artificial neural networks were shown to be quite efficacious in modelling certain cognitive phenomena that had been problematic to implement with symbolic computational architecture.

The operations of artificial neural networks are designed to mimic the neural circuitry of the brain – they are often referred to as implementing 'brain style' processing. As such, it may aid your understanding of this chapter to first revisit the discussion of the operations of neurons in Chapter 4.

In this chapter we are going to develop a sound understanding of the operations of artificial neural networks and their utility in modelling cognitive functions. We'll begin by describing the basic connectionist architecture and explaining how this differs from symbolic computational architecture.

19.1 CONNECTIONIST ARCHITECTURE

Classical symbolic computational architecture – which we described at length in Chapters 7 to 9 and have seen many examples of since – admits of the following essential features.

Firstly, there is only one processor in the architecture – a central processing unit (CPU) which processes program instructions. Secondly, the CPU carries out these instructions serially – one after the other. Thirdly, the CPU addresses and operates on localised register contents.

Connectionist architecture, on the other hand, is crucially distinct with respect to each of these features. Connectionist networks are composed of a (typically large) number of simple processing units (nodes) which operate in parallel rather than serially. Content in connectionist networks is not local and addressable, but distributed across numerous nodes and encoded as a pattern of connections.

The basic elements of an artificial neural network are simple processing units which are designed to emulate the operations of individual neurons. These units are functionally organised in layers – there will be an input layer of nodes and an output layer of nodes. There will typically also be a 'hidden' layer of nodes – these are neither input nor output units but serve to mediate between these layers.

As you have no doubt determined, nodes are connected to each other. Precisely how they are interconnected defines various architectural variations which needn't concern us much here. In networks of interesting complexity, each node will be connected to a large number of other nodes – just as individual neurons are connected to large numbers of other neurons. The simplest type of connectionist architecture (or the most complex depending on how you look at it) is such that every node is connected to every other node in the network.

Information processing in artificial neural networks is achieved through the propagation of *activation* along the connections through the network. Each node in the network has a level of activation which is influenced by the activation it receives from other nodes which are connected to it.

We're going to make some simplifying assumptions here about activation. Firstly, we're going to assume that at each time step, the activation of a node is entirely determined by the activation it receives along its incoming (afferent) connections (rather than consider a more complicated function which also takes into account the antecedent level of activation of the node from the previous time step).

Connections between nodes can be either *excitatory* or *inhibitory* and this is represented by assigning a *weight* – a positive or negative numerical value – to each connection. Excitatory connections – which are positively weighted – will *excite* (increase the level of activation of) the node they are connected to. Inhibitory connections – which are negatively weighted – will *inhibit* (decrease the level of activation of) the node to which they are connected.

Each node in the network, you will recall, is a simple processing unit. These nodes implement two functions – an *activation* function and a *transfer* function.

The *activation* function determines whether or not a node will *fire* based on its level of activation at that time step. We're only going to consider the simplest of activation functions – a *threshold* function. Nodes with a threshold activation function will fire iff their level of activation at that time step is above some threshold value assigned to the node. If a node fires, it passes activation along each of its outgoing (efferent) connections to other nodes, otherwise no activation propagates through that node.

The *transfer* function determines how a node updates its level of activation based on the activation it receives along its afferent connections. Again, we're only going to consider the simplest of transfer functions – a *weighted sum* function. To determine the level of activation of a node with a weighted sum transfer function, we simply take the sum of the values of the afferent connection weights.

19.2 SIMPLE ARTIFICIAL NEURAL NETWORKS

Let's take a look at some basic examples to exemplify these operations. To keep things simple, I'm going to use integers for connection weights and threshold values. Figure 19.1 depicts the simplest artificial neural network that does something interesting.

This network has two input nodes (A and B) and one output node (C). We're interested in whether or not the output node will fire (although its efferent connection is not afferent to any other node). The input nodes we can imagine as detectors of some kind. They are set to fire if some environmental condition is met – perhaps if a light is on or if a switch is in a particular position.

The two connections in the network are both excitatory and equally weighted. If A fires it excites C and if B fires it excites C. The

A	B	A & B
T	T	T
T	F	F
F	T	F
F	F	F

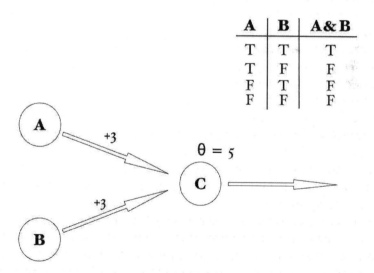

Figure 19.1 Computing AND.

threshold value θ of C is such that it will only fire if *both* A and B fire. If A alone fires, then the activation value of C will be 3 which is below the threshold value of 5. Similarly if B alone fires. If, however, they both fire, then the weighted sum transfer function tells us that the activation value of C will be the sum of the values of the afferent connection weights, which in this case will be 6. This activation value of 6 is higher than the threshold value of 5 assigned to C, so it will fire in accordance with its threshold activation function.

This network serves as a *logic gate*. It computes the binary logical truth function of conjunction. The output unit fires iff both A *and* B fire.

Exercise 19.1

(a) How might we modify the network depicted in Figure 19.1 such that it computes the binary logical function of *disjunction* – i.e. so that the output node fires iff either A *or* B (or both) fire.

(b) Design a network with two input nodes and one output node such that the output node will fire if either of the input nodes fire but *not* if they both fire.

If you succeeded in completing Exercise 19.1(b) then you have designed a network which computes the binary logical truth function of *exclusive disjunction*. This is not a straightforward exercise since we need to do two things that are new to us. One is to assign inhibitory weights; the other is to add a hidden unit to the network between the input and output units. The solution is depicted in Figure 19.2.

If we disregard node D in Figure 19.2 for a moment, then we have the solution to Exercise 19.1(a). All we needed to do was lower the threshold value of C to a value below either of the afferent connection weights. (Alternatively we could have raised both connection weights to a value above the threshold.) In order to prevent C from firing when *both* A and B fire, however, we need to add node D.

Node D will fire iff both A and B fire and will inhibit the activation of C to prevent it from firing. If A alone fires, then the activation value of D will be below the threshold value and it will not fire, but the activation value of C will be above its threshold and it will fire. Similarly if B alone fires. If A and B *both* fire, however, then D will fire, as its activation will be above its threshold value. The activation value of C will be the sum of its afferent connection weights $(3 + 3 + -5 = 1)$

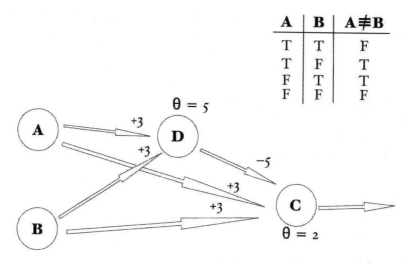

A	B	A≢B
T	T	F
T	F	T
F	T	T
F	F	F

Figure 19.2   Computing XOR.

which is below its threshold value so, by virtue of being inhibited by D, it will not fire. To recap, C will fire if either A fires or B fires but *not* if they both fire, *quod erat demonstrandum*.

19.3 SYNTHESISING SPEECH

In this section we're going to design an artificial neural network to function as an English speech synthesiser. This is a nice example of the kind of contextually sensitive processing tasks at which connectionist networks excel.

English orthography is not phonemic – it does not admit of a regular mapping onto English phonemes. Unlike Japanese, for instance, which is such that the pronunciation of any given grapheme is contextually invariant, the pronunciation of a given English grapheme is dependent on its orthographic context. In other words, English is not 'pronounced as it is written', to speak loosely. Rather, the sound that a given letter stands for is contextually dependent – one and the same letter can stand for different sounds and the same sound can be represented by various letters, depending on the spelling context.

Consequently, designing a speech synthesiser for English involves, *inter alia*, implementing the contextually sensitive processing task of converting orthographic representations to phonemic representations. We're going to begin constructing an artificial neural network for implementing this conversion of orthographic input to phonemic output.

If we want our network to be contextually sensitive, we're obviously going to need context at the input layer to be sensitive to. We're going to achieve this by organising the input layer into five pools of nodes. Each pool will contain twenty-seven nodes, representing a full set of letter detectors – one for each letter of the alphabet plus one for the space (we're going to ignore punctuation here to keep things simple).

These input pools are going to be organised such that each pool is directed on a different letter position of the text string input. At each time step, one pool will be directed at a target letter position. The other four pools will be organised such that one pool is directed at each of the two letter positions on either side of the target position. This will allow the network to make a contextual determination of which phoneme a particular letter stands for given the surrounding orthographic context (the two letters either side of the target letter).

At the first time step, the first letter in the text string is placed in the target position. At each subsequent time step, the text string is advanced such that the next letter in the string is in the target position.

The output layer of our speech synthesising network will consist of an output node for each phoneme, so if we are considering Australian English there will be forty-four output nodes. To keep things simple here, we're going to consider just one phoneme whose pronunciation is invariant across English dialects: /s/ – the word final sound in 'kiss' and 'this'.

We can help ourselves to as many hidden units as we require in order to match inputs to outputs correctly. We're going to set threshold values and connection weights such that our network makes correct determinations concerning whether or not /s/ should be pronounced for the following test set of words: this, gas, wish, shy, kiss, passive, asia, asian, asiatic, is, as, ice, justice, service.

The first thing to do is to accommodate the standard case. When we see the letter 's', it is usually the case that the phoneme /s/ should be produced. So the first thing we'll do is to connect the 's' detector in the input pool for the target letter position directly to the output unit representing the phoneme /s/ such that if the letter 's' is detected in the target position, the /s/ unit will fire unless otherwise inhibited (see Figure 19.3).

Our network will now perform correctly with respect to the first two words in our test set – 'this' and 'gas'. When the 's' in each word reaches the target position, the /s/ unit will fire, as it should.

There are, however, numerous words in which the letter 's' does not represent the phoneme /s/. Our nascent speech synthesising network

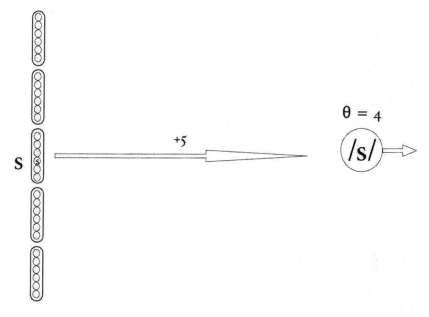

Figure 19.3   The standard case.

will currently make incorrect determinations with respect to the remainder of the words in our test list. Our next task then is to design hidden units which detect contexts in which the letter 's' appears but the phoneme /s/ should not be produced and use these hidden units to inhibit the activation of the output unit accordingly.

We want the contexts represented in the hidden layer to be as general as possible so as to accommodate the maximum number of cases. It is almost always the case in English – with the exception of some proper names and compound words – that when a letter 's' is followed by a letter 'h' it is not pronounced as /s/. The first hidden unit we will add to the network will detect just such contexts and inhibit the output unit (see Figure 19.4).

Now, when our network is presented with either of the next two words in our test set – 'wish' or 'shy' – the hidden unit we added will inhibit the /s/ unit such that it will not fire, as required. As should be clear, the network now considers any orthographic context in which 's' is followed by 'h' to be a context in which /s/ is not produced.

Another context which exhibits similar regularity is one in which 's' is followed by another 's', such as our test words 'kiss' and 'passive'. In such cases the phoneme /s/ is produced, but only once. As it stands, our network will determine that /s/ should be pronounced twice as the output unit will fire when each 's' is in the target position.

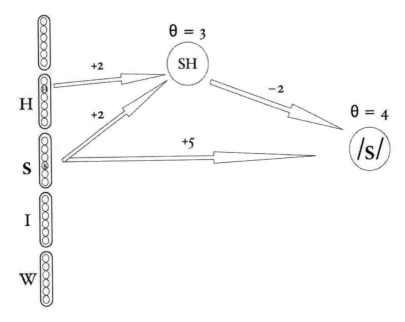

Figure 19.4 Context sensitivity.

Consequently, we need to add another hidden unit to accommodate double 's' contexts (see Figure 19.5).

Note that we could have implemented a slightly different solution to the double 's' problem. It makes no difference which letter 's' we inhibit the production of the phoneme /s/ for, so long as it is only produced once. As such, we could just as well have inhibited the firing of the output node for an 's' with a following 's' rather than a preceding one.

The next three words in our test set – 'asia', 'asian' and 'asiatic' – can all be accommodated with the addition of a single hidden unit. No word in English with the letter combination 'asia' is such that the 's' is pronounced as /s/. We can therefore add a hidden node to the network which detects the context 'asia' and inhibits the firing of the output unit.

Exercise 19.2

Augment the network by adding a hidden unit and setting weights and thresholds to detect the context 'asia' and inhibit the firing of the output unit.

Accommodating each of the following two words in the test set – 'is' and 'as' – requires more specificity. While we want our context

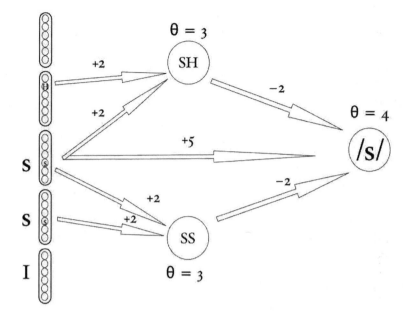

Figure 19.5   Context sensitivity.

detectors to be as general as possible, these two cases don't admit of contextual generalisation. We might be tempted, for instance, to inhibit the firing of the output unit for any context in which 'is' appears at the end of a word. While this would then also accommodate 'his', the network would subsequently make an incorrect determination when presented with 'this'. Similarly with 'as', 'has' and 'gas'.

Consequently, we need to add hidden units which detect just the words 'is' and 'as'. This is where the utility of having a node in each input pool for detecting a space becomes apparent. Detecting just the word 'is' involves detecting the context '_is_', where '_' represents a space.

Exercise 19.3

Augment the network by adding hidden units and setting weights and thresholds to detect the contexts '_as_' and '_is_' and inhibit the firing of the output unit.

As well as there being numerous contexts with the letter 's' such that the phoneme /s/ should not be produced, there are also numerous ortho-graphic contexts such that the phoneme /s/ should be produced which do not contain 's'. The last two words in our test set are just such cases.

Almost every English word containing the letter combination 'ice' is such that the 'c' is pronounced as /s/. The simplest way to accommodate the remaining words in our test set is to add a hidden node which detects this context and excites the output unit such that it will fire, as required.

Exercise 19.4

Augment the network by adding a hidden unit and setting weights and thresholds to accommodate the context 'ice'.

Exercise 19.5

Augment the network further so as to also accommodate the following words: precise, recede, perceive, receive, precipitate, reception, recipe.

Exercise 19.6 (Challenge)

(a) Try to accommodate as many of the following words as possible, without causing the network to make incorrect determinations with respect to any of the words in our original test set or the extended set of Exercise 19.5: ease, lease, please, peace, grease, guise, reprise, practise, practice, his, has, mission, passion.

(b) What is preventing us from accommodating all of these words? How might we extend our network architecture to improve this?

19.4 LEARNING

You should now have a sense of precisely how complex a processing task it is to convert English orthography to phonemics. We've considered only one phoneme and only a tiny fraction of relevant cases and even this quickly became quite a complex task.

It also turned out that our network architecture was insufficiently complex to accommodate even a small test set of words. While we made provisions for some context at the input layer, we didn't allow for sufficient context to make accurate determinations with respect to the full range of possible contexts in English.

Designing a correctly functioning speech synthesising network for English in its entirety by designing hidden units and setting weights and thresholds by hand would be a highly labour-intensive exercise. Fortunately, however, artificial neural networks also excel at *learning* to match inputs to outputs.

While the network of the previous section is a nice example of the operations of artificial neural networks, we would not ordinarily construct a network of any interesting complexity in this fashion. Specifying the function of nodes in the hidden layer the way we've done belies the appellation 'hidden'. Typically, threshold values and connection weights for nodes in the hidden layer are determined by *training* the network.

There are numerous training methodologies for artificial neural networks. A common methodology involves employing a *backpropagation* algorithm to revise connection weights and threshold values based on a generated error value.

Backpropagation of error is a supervised training methodology, which means that we have an antecedent determination of how inputs should be correctly mapped to outputs – e.g. in our speech synthesising network we want the output unit for a given phoneme to fire always and only when that phoneme should be produced given the orthographic input context.

Training an artificial neural network to function as a speech synthesiser using backpropagation of error would involve the following. We'd begin with the same input pools and output nodes we described in the previous section (although we'll want a wider input window – more input pools – to provide more context). We'd then add a large number of hidden nodes and connect every input node to every hidden node and every hidden node to every output node (giving us a maximally interconnected *feedforward* architecture).

The goal is to get the network to perform correctly on a training set of data – such as our test sets of words. We begin by simply assigning small random values to connection weights and thresholds and testing the resulting performance. Initially, the network will perform very poorly – failing to correctly match inputs to outputs – as we'd expect. We then generate an error value which indexes how far the network has deviated from the correct mapping. This error value is then propagated back through the network and adjustments are made to weights and thresholds according to our backpropagation algorithm.

The technical details needn't concern us here as the calculus involved is moderately complex. For our purposes, a conceptual understanding of the training process suffices. After cycling the network and backpropagating the error many times, the network will eventually converge on a state which facilitates a correct mapping from inputs to outputs on the training set of data.

If our training set of data is sufficiently large and sufficiently representative so as to adequately characterise the relevant space of possible mappings, the network's correct performance on the training set should generalise to novel inputs. In the case of our speech synthesising network, we can say that the network has 'learned' to produce correct phonemic transcriptions of orthographic representations.

19.5 PATTERN RECOGNITION

In our described example case of training an artificial neural network to translate orthography to phonemics, the network learns how to map orthographic contexts to phonemes by learning to recognise certain patterns.

During the training process, the network extracts patterns from the input data and these are encoded as the connection weights and unit thresholds among numerous nodes. In other words, various patterns detected in the training data are encoded in the network as distributed representations.

Although I've not demonstrated it here, artificial neural networks are able to recognise (token the representations for) learned patterns, even given partial or noisy input. It is this ability to extract and encode patterns occurring in data sets and then recognise these patterns when they occur in subsequent inputs – even when the input is less than perfect – that makes artificial neural networks well suited to modelling a range of cognitive phenomena.

In our example network, these patterns were particular orthographic contexts; however, they could be any kind of pattern, depending on the input. Patterns arise in all kinds of environmental stimuli and the human capacity to be sensitive to the occurrence of particular patterns is fundamentally implicated in a broad range of cognitive capacities.

Our rational capacity, in particular, is contingent on this ability. Very often our intuitive reasoning involves analogical comparison to structurally similar situations in our past experience – this is a type of pattern matching. Even deliberately following formal methods of reasoning, you will recall, required us to be able to discern logical forms and match these to the antecedent forms of logical rules.

These properties of artificial neural networks – their contextual sensitivity and amenability to various training methodologies – bode well for their successful deployment in artificial intelligence projects. They also enjoy other relevant advantages.

Artificial neural networks are readily scalable. Given, for instance, the speech synthesising network fragment from section 19.3, it is a simple matter to augment the network to make determinations concerning the production of other phonemes, or to widen the input layer to take a broader context into account. This is aided by the fact that units in the hidden layer can serve several processing functions – for example, a collection of units that detect a certain pattern can simultaneously excite some output units while inhibiting others.

Artificial neural networks are also – in principle – amenable to interconnection. Consider the word 'read'. Orthographic context alone is insufficient to determine which vowel phoneme should be produced when uttering this word. We also need to determine its *syntactic* context, since this is what tells us which vowel to produce. Consequently, if we want our speech synthesising network to perform correctly on this and similar cases, it will need to interoperate with a network making syntactic classifications.

Similarly, if we want a completed speech synthesising network that can produce 'natural' sounding speech, we need to apply phonetic realisation rules to the phonemic output. We also, crucially, need to make various semantic and pragmatic determinations in order to establish the intonation contours of utterances. These are extraordinarily difficult problems which might be solvable by a number of specialised neural networks interoperating in parallel to subserve linguistic production.

Finally, artificial neural networks typically exhibit graceful degradation, in much the same way human brains do. Removing a single element from a register machine – a register or a line of code – is usually sufficient to break it completely. Artificial neural networks, on the other hand, are more robust to damage. Removing a small number of elements may have little or no effect. Detrimental effects, when they arise as a result of further lesioning, may well be ameliorated by retraining the lesioned network such that it recovers its functions – just as stroke patients relearn cognitive functions.

19.6 TWO PARADIGMS?

Although I've described the symbolic and connectionist approaches to artificial intelligence as fundamentally distinct – and, by implication, incommensurate – *paradigms*, it may well be the case that these views concerning information processing merely engage at different levels of description.

The connectionist paradigm is often referred to as the *sub-symbolic* paradigm, implying that it engages at a lower level of description than symbol systems. Alternatively, it may well be the case that in human cognition, certain kinds of low-level symbolic processing subserve higher-level connectionist processing.

It seems, prima facie, that the operations of artificial neural networks (at least as we've described them here) are entirely effective. Hence, by the Church-Turing thesis, they are register machine computable. Certainly transfer functions and activation functions are algorithmic and it seems we can approximate parallel processing with stepwise serial processing, so perhaps connectionism simply reduces to symbolic processing.

On the other hand, we have seen how to construct logic gates with artificial neural networks. Computers as we standardly know them are essentially constructed from logic gates, so perhaps symbol systems simply reduce to connectionist processing.

In practice, symbolic models and artificial neural network models are not at all radically incommensurate, since artificial neural networks are simulated on symbol systems architecture. Recent advances in the nascent and burgeoning field of neural bioengineering, however, are taking the symbol systems out of the equation. Biological neural networks, constructed from actual neurons, have been shown to exhibit many of the features of artificial neural network models – including the capacity to be trained to implement particular complex functions, such as proficiently operating a flight simulator.

19.7 IT'S ONLY A MODEL

The introduction to artificial neural networks in this chapter has been very basic indeed. We've considered only the simplest kinds of networks and functions in order to avoid unnecessary mathematical complexity. I'd strongly recommend that the interested reader continue their investigations with the suggestions for further reading as a guide. A proper introduction to artificial neural networks requires a dedicated textbook.

Even in all their sophistication and complexity, artificial neural network models remain gross simplifications of the biological neural activity which they seek to model. In particular they fail to take into account the global and analogue effects of neurotransmitters and this has profound implications for the possibility of modelling a number of mental phenomena, including (crucially) attention and the emotions.

As we learn more about the brain, however, we may be able to develop yet more sophisticated models which implement the neurobiological principles we uncover. It will be particularly interesting to see whether developments in neural bioengineering over the next decade provide empirical fodder for computational neural modelling.

CHAPTER 20

MINDS AND COMPUTERS

We have now learned a lot about minds, having surveyed the space of available philosophical theories of mind and considered the advantages and disadvantages of each theory.

We've also learned a lot about computers, having developed a rigorous technical account of precisely what a computer is and practised the fundamentals of computer programming.

We've seen how we might employ symbol systems to implement a number of functions implicated in cognition – particularly with respect to the rational and linguistic capacity.

Along the way we've learned some basic functional neuroanatomy, a little formal logic, a sprinkling of linguistics and, as well as briefly touching on modern cognitive psychology, we've learned about the early history of empirical psychology.

Finally, we've looked at some simple artificial neural networks and have seen how we might employ such connectionist networks in modelling cognitive phenomena – again, with particular respect to the rational and linguistic capacity.

All this has been in the service of an interdisciplinary examination of the tenability of the project of strong artificial intelligence.

In this final chapter, I want to just briefly touch on some of the philosophically 'hard' problems related to artificial intelligence – namely those associated with consciousness, personal identity and the emotions.

20.1 CONSCIOUSNESS

Although I've helped myself in places to an intuitive distinction between the mental processes we are consciously aware of and those which occur below the level of consciousness, I've not said much at all about consciousness per se.

It is an ineliminable – but perhaps not irreducible – fact about human mentality that we have *consciousness*. The word 'conscious', however, is used in many ways.

Sometimes it is used to refer to our awareness of certain events or processes that we are *conscious of*. Sometimes it is used to refer to our awareness of our self and the distinction between our self and the rest of the world – our *self consciousness*. Sometimes it used merely to distinguish between our waking states and sleeping – or *unconscious* – states.

In certain religions, to have *consciousness* means to be ensouled. In psychoanalytic theory, the *conscious* mind is commonly distinguished from the *subconscious* mind and these are typically held to be in all kinds of tensions that only psychoanalysis can resolve.

More philosophically, being conscious involves having the capacity for subjective experience and for having the associated privileged first-person qualities of experience – *qualia*. It is also strongly associated with the capacity for developing representational states with intentional content.

It is less than clear if there is a single overarching 'problem of consciousness' or a number of relevant problems – perhaps 'easier' and 'harder' problems – although David Chalmers has done much to disambiguate senses and tease apart the philosophical issues involved.

Consciousness is currently the hot topic in philosophy of mind with dedicated research centres arising to investigate the phenomenon. These research centres are engaged in the kind of interdisciplinary analysis we have conducted in this volume, with a strong focus on determining precisely what the relevant philosophical questions are and how one might go about answering them.

Philosophically advanced readers would be well advised to follow the suggestions for further reading to develop their understanding of this challenging, engaging and developing area of philosophy.

20.2 PERSONAL IDENTITY

On any given day, I clearly differ in a number of ways from the way I was the day before since I will have a number of different properties. I will have a distinct spatiotemporal location, I may have altered or augmented beliefs, I will have extra memories, small bits of my body – skin, hair, fingernails and the like – will have been lost and new bits will have grown, and so on.

However, despite these numerous distinct properties from day to day and year to year, I am always the *same person* – I have a unique personal identity which endures through numerous changes in my spatiotemporal, psychological and material properties. Although I am qualitatively distinct from day to day, I am numerically identical – i.e. one and the same person.

It is not difficult to problematise the notion of enduring personal identity. Regardless of one's preferred criteria for the persistence of personal identity through time, it seems we can come up with problem cases.

It is common to privilege psychological continuity as a criterion for the persistence of personal identity. Psychological continuity, however, doesn't seem to be a *necessary* condition for the persistence of personal identity since I can imagine having total amnesia such that my psychological states are radically discontinuous with past psychological states. Intuitively though, I would still be the *same person* – I would have just lost my memory. There is a response or two available here but I leave this up to the reader.

Nor does the psychological continuity criterion seem *sufficient* for the persistence of personal identity. Suppose I step into a matter teleportation device and, through some mishap, I am reassembled twice at the other end. Both of the beings who step out of the matter transporter are psychologically continuous with me – at least at the instant they step out – so, if psychological continuity is a sufficient criterion for the persistence of personal identity, both must be the same person as me. Again, I leave responses to the reader.

Now suppose that computationalism is true. This means that, in principle, I could replicate your mind exactly with computational hardware other than your brain. Suppose I have a computational device sufficiently powerful to run your [MIND] equally as well as your brain does. Further suppose that one night while you are sleeping, I use a fancy new scanner and some innovative surgical techniques to scan your [MIND], replicate it in my computational device and then replace your brain with the new computational device without your ever being aware.

What do your intuitions tell you in this case? Are you still the same person? If you think not then modify the example so that on the first night, I replace just one of your neurons with an artificial neuron. On the next night, I replace ten. On the next night, I replace a hundred, then a thousand, then a million, and so on – all without your ever being aware. Unless you're prepared to indicate which number of replaced neurons constitutes a change in your personal identity, it seems you must be committed to being the same person at the end of this process.

If you think that you are the same person with the alternative computational device replacing your brain, then modify the example so that I merely scan your [MIND] and then place the computational device implementing it into an android body, leaving you just as you are. What do your intuitions tell you now? What obligations do we have, if any, to the android body with your [MIND]?

I have included a couple of articles in the suggestions for further reading which problematise personal identity while remaining accessible and entertaining to the introductory reader.

20.3 EMOTIONS

It is generally considered that to lack the capacity for emotional states and responses is to lack one of the requirements for having a mind in the sense that we have minds. A deficit in emotional behaviour is one of the characteristic symptoms of certain psychopathologies and, while we hold such people to have minds, we believe their minds to be importantly qualitatively distinct from our own.

It is almost always emotional engagement that is used to blur the line between humans and artificial intelligence in science fiction. Think of the endings of the movies *Bladerunner, Terminator II* and *I, Robot* as examples. In each case, we are led to being emotionally disposed towards the artificially intelligent protagonist by virtue of coming to believe that it is capable of having emotions.

Emotion is one of the least well understood aspects of mentality. We know that certain emotions are correlated with, and can be stimulated by, certain neurotransmitter combinations, but our understanding of these processes is sketchy at best. We also know that damage to certain localised areas of the brain can result in characteristic emotional deficits.

One of the particularly interesting things we know about emotion is that emotional engagement is strongly tied to episodic memory. It is manifestly the case that we are much more likely to remember events which invoked in us a strong emotional response. Furthermore, we know the limbic system of the brain to be implicated in both emotion and memory.

It is intuitively clear, particularly on reflection of the science fiction examples I mentioned, that we are much more likely to believe that an artificial intelligence with the full range of human emotional responses qualifies as having a mind in the same sense that we have minds. However, the problem of the inaccessibility of another mind's qualitative aspects of experience arises again here.

If an artificial intelligence displayed the standard range of human emotional responses but these were just outward displays which didn't *feel like* anything to the artificial intelligence, would we still attribute to it the robust notion of having a mind? If not, then why do we attribute having a mind to other people when we all we are able to discern about their emotional states is their observable behaviour?

As always, it is less than clear what one should say about qualia and I leave this to the reader to consider.

20.4 COMPUTERS WITH MINDS

Now that we've reached the end of the book, it is time to reflect on what determinations we are able to make concerning the possibility of artificial intelligence.

We haven't seen anything here which leads us to believe that strong artificial intelligence is *impossible*, although we have seen some entry points for mounting such arguments. Prima facie, with a concession to the potential determinations of further philosophical investigation, it seems that it may well be possible to design a computer which has a mind in the sense that we have minds.

We have, however, seen that our current best computational models of cognition are still woefully inadequate, but we hold out hope that advances in neuroscience may provide us with technical understandings of the biological processes subserving cognition which will lead to richer conceptual understandings and, ultimately, successful strong artificial intelligence projects.

We have managed to impose some putatively necessary conditions on the development of artificial intelligence. In Chapter 17, we argued that embodied experience was a necessary condition for the development of semantics which, in turn, are necessary for having a mind.

Consequently, if we want to develop an artificial intelligence it must, in the first instance, be connected to the external world in the relevant ways. In other words, it must enjoy sensory apparatus which mediate the relations between it and the external world. Furthermore, our embryonic artificial intelligence must then be able to gather a weight of experience, through which it will be conferred with mental representations.

Given our current conceptual understanding of the mind and technical understanding of the computational wetware of the brain which gives rise to it, by far the simplest way to create something which has the capacity for embodied experience and which is *ceteris paribus* guaranteed to develop a mind in the same sense that we have a mind is still the old-fashioned biological way – to create and raise a human being.

APPENDIX I: SUGGESTIONS FOR FURTHER READING

CHAPTER 2

Campbell, K. *Body and Mind*. London: Macmillan, 1970, ch. 3.

Churchland, P. *Matter and Consciousness*. Cambridge, MA: MIT Press, 1988, ch. 2.

Descartes, *Meditations on First Philosophy*, trans. J. Cottingham. Cambridge: Cambridge University Press, 1986, pp. 50–6.

CHAPTER 3

Campbell, K. *Body and Mind*. London: Macmillan, 1970, ch. 4.

Gardner, H. *The Mind's New Science*. New York: Basic Books, 1985, pp. 98–114.

Ryle, G. *The Concept of Mind*. Harmondsworth: Penguin, 1973, pp. 13–25.

Schultz, D. *A History of Modern Psychology*. New York: Academic Press, 1975, chs 3, 4, 5, 10, 11.

CHAPTER 4

Barr, M. *The Human Nervous System: An Anatomic Viewpoint*, 3rd edn. Hagerstown, MD: Harper & Row, 1974.

Diamond, M. C. et al. *The Human Brain Coloring Book*. New York: Harper Collins, 1985.

Gregory, R. (ed.) *The Oxford Companion to the Mind*, Oxford: Oxford University Press, 1987, pp. 511–60.

CHAPTER 5

Armstrong, D. *A Materialist Theory of the Mind*. London: Routledge & Kegan Paul, 1968.

Jackson, F. 'Epiphenomenal Qualia', *Philosophical Quarterly*, 32 (1982), pp. 127–36.

Nagel, T. 'What Is It Like to Be a Bat?', *Philosophical Review*, 83 (1974), pp. 435–50.

Place, U. T. 'Is Consciousness a Brain Process?', *British Journal of Psychology*, 47 (1956), pp. 44–50.

Smart, J. J. C. 'Sensations and Brain Processes', *Philosophical Review*, 68 (1959), pp. 141–56.

CHAPTER 6

Block, N. 'Troubles with Functionalism', reprinted in Block, N. (ed.) *Readings in Philosophy of Psychology*, Vol. 1. Cambridge, MA: Harvard University Press, 1980.

Guttenplan, (ed.) *A Companion to the Philosophy of Mind*. Oxford: Blackwell, 1994, pp. 317–32.

CHAPTERS 7–9

Church, A. 'An Unsolvable Problem of Elementary Number Theory', *American Journal of Mathematics*, 58 (1936), pp. 345–63.

Hofstadter, D. *Gödel, Escher, Bach: An Eternal Golden Braid*, 20th Anniversary edn. London: Penguin, 2000, pp. 33–41.

Jeffrey, R. *Formal Logic: Its Scope and Limits*, 3rd edn. New York: McGraw-Hill, 1991, pp. 100–2.

Turing, A. M. 'On Computable Numbers, with an Application to the Entscheidungsproblem', *Proc. London Math. Soc.*, 42 (1937), pp. 230–65.

CHAPTER 10

Scheutz, M. (ed.) *Computationalism: New Directions*, Cambridge, MA: MIT Press, 2002, ch. 1.

Turing, A. M. 'Computing Machinery and Intelligence', *Mind*, 59 (1950), pp. 433–60.

CHAPTERS 11–13

Copeland, J. *Artifical Intelligence: A Philosophical Introduction*. Oxford: Blackwell, 1993, ch. 4.

Haugeland, J. *Artificial Intelligence: The Very Idea*. Cambridge, MA: MIT Press, 1989, ch. 2.

Newell, A. and Simon, H. 'Computer Science as Empirical Enquiry: Symbols and Search', *Communications of the ACM*, 19 (1976), pp. 113–26.

Pinker, S. *How The Mind Works.* London: Allen Lane, 1998, ch. 2.

CHAPTER 14

Chomsky, N. *Aspects of the Theory of Syntax.* Cambridge, MA: MIT Press, 1965.

Haegeman, L. *Government and Binding Theory.* Oxford: Blackwell, 1991.

CHAPTER 15

Cohen, L. J. 'Can Human Irrationality be Experimentally Demonstrated?', *Behavioural and Brain Sciences*, 4 (1981), pp. 317–30.

Gardner, H. *The Mind's New Science.* New York: Basic Books, 1985, ch. 13.

Johnson-Laird, P. N. *Mental Models: Towards a Cognitive Science of Language, Inference and Consciousness.* Cambridge, MA: Harvard University Press, 1983.

Kahneman, D., Slovic, P. and Tversky, A. (eds) *Judgment under Uncertainty: Heuristics and Biases.* Cambridge: Cambridge University Press, 1982.

Wason, P. C. 'Natural and Contrived Experience in a Reasoning Problem', *Quarterly Journal of Experimental Psychology*, 23 (1971), pp. 63–71.

CHAPTER 16

Crystal, D. *The Cambridge Encyclopedia of the English Language.* Cambridge: Cambridge University Press, 1995, ch. 17.

Fromkin, V., Rodman, R. et al. *An Introduction to Language*, 2nd edn. Sydney: Holt, Rinehart & Winston, 1990, chs 2, 3.

CHAPTER 17

Searle, J. 'Minds, Brains and Programs', *Behavioural and Brain Sciences*, 3 (1980), pp. 417–57.

CHAPTER 18

Fodor, J. *Concepts: Where Cognitive Science Went Wrong*. Oxford: Clarendon, 1998.

Putnam, H. *Reason, Truth and History*. Cambridge: Cambridge University Press, 1981, ch. 1.

Smith, E. and Medin, D. *Categories and Concepts*. Cambridge, MA: MIT Press, 1981.

CHAPTER 19

Dennis, S. and McAuley, D. *Connectionist Models of Cognition*. Online text, available at the time of writing at: http://lsa.colorado.edu/~simon/cmc/index.html.

Rumelhart, D. and McClelland, J. et al. *Parallel Distributed Processing: Explorations in the Microstructure of Cognition. Vol. 1: Foundations*. Cambridge, MA: MIT Press, 1986, chs 1, 2, 3, 5, 8.

CHAPTER 20

Chalmers, D. *The Conscious Mind*. New York: Oxford University Press, 1996.

Dennett, D. *Brainstorms: Philosophical Essays on Mind and Psychology*. Cambridge, MA: MIT Press, 1981.

Hofstadter, D. and Dennett, D. (eds) *The Mind's I*. New York: Basic Books, 1981, ch. 26.

APPENDIX II: GLOSSARY OF TERMS

a fortiori even more strongly, by the same reasoning. 'Brunswick is in Melbourne, therefore it is in Victoria; *a fortiori*, it is in Australia.'

activation function one of the two functions implemented by each processing node of an *artificial neural network* which determines, given a level of activation, whether or not the node should fire.

adicity the number of inputs which a *function* takes.

afferent incoming (signal conduction).

algorithm another name for an *effective* procedure – a systematic method which requires no understanding for its execution and which can be completed in finite time.

allophone a phonetic variant of a *phoneme* whose context-dependent production is determined by *phonetic realisation rules*.

amphiboly a property of certain *syntactic* structures such that they admit of multiple *semantic* interpretations. 'I saw the man on the hill with the telescope.'

anomalous monism the view that certain physical states have irreducibly non physical properties.

antecedent the left-hand side of a *conditional*.

aphasia a language deficit. Common forms include Broca's aphasia, Wernicke's aphasia and conduction aphasia.

artificial intelligence, strong the contentious view that it is possible to develop artefacts which have minds in the sense that we have minds.

artificial intelligence, weak the uncontentious view that it is possible to create artefacts which are able to implement certain functions which are held to be (weakly) constitutive of intelligence. Often used to sell white goods.

artificial neural network a structure of weighted connections between simple processing units – each of which implements a *transfer function* and an *activation function* – the propagation of

211

activation through which constitutes parallel distributed processing. Artificial neural networks are modelled on 'brain style' information processing.

behaviourism, philosophical (analytic)   a semantically reductive and ontologically eliminative theory of mind, according to which our mental state terms don't actually pick anything out but are simply useful shorthand locutions for referring to complex sets of dispositions to behave.

behaviourism, psychological   a methodological view concerning how psychological investigation should proceed, according to which psychology should treat exclusively of observable behaviour. Influenced by *positivism*.

causal theory (of mind)   the view that mental states are characterised exclusively in terms of their causal role with respect to mediating relations between stimulus and behaviour.

ceteris paribus   all other things being equal.

Chinese room   a thought experiment which seeks to show that no amount of *syntactic* manipulation, in isolation from the external world, is sufficient to generate *semantics*.

cognitive architecture   the structure and nature of the information processing systems of a cognitive agent – the organisational and implementation features of the information processing hardware of a cognitive agent.

computation   the sequence of operations of a *register machine*.

computationalism   the philosophical theory of mind according to which minds are akin to the software being run on the hardware, or *wetware*, of the brain – a kind of *functionalism* which allows for the possibility of computational *artificial intelligence*.

computer   the physical implementation of – or, more accurately, approximation to – a *universal machine*.

conditional   a statement of the form 'If . . . then . . .'.

conjunction   a logical operation represented by the natural language term 'and'.

consequent   the right-hand side of a *conditional*.

counter-example   an interpretation which shows that an inference form is not *valid* by showing that it is possible for all the premises to be true while the conclusion is false.

decidable   a property of some *formal systems* such that there is an *effective* procedure for determining, of any given *state* of the system, whether or not it is *generated*.

deduction   an inference which appeals to some logical principle, such as *modus ponens*.

derivation a proof that a particular *state* of some *formal system* is *generated* which demonstrates the required sequence of applications of the *rules* of the system.

deterministic a property of some *formal systems* such that, at most, one *rule* will apply to any given *state* and in only one way.

diphthong a *sonorant phoneme* whose production requires movement of the tongue from one cardinal vowel position to another.

distinctive contrast two *phones* display *distinctive contrast* – and are thereby distinct *phonemes* – *iff* the substitution of one *phone* for another results in a change of meaning.

dualism, Cartesian the philosophical theory of mind – otherwise known as interactionist dualism, according to which the mind and body are composed of distinct substances which interact with each other.

dualism, substance an *ontological* view according to which the universe is composed of two distinct substances – physical and non-physical, or material and immaterial.

effectivity a property of certain procedures – otherwise known as *algorithms* – such that they can be carried out systematically, in finite time, without the need for understanding. Effectivity constrains the *states* and *rules* of *formal systems*.

efferent outgoing (signal conduction).

epiphenomenalism the philosophical theory of mind according to which mental states are non-physical but causally inert – mere epiphenomena of certain physical processes.

expert system a particular kind of *formal system* used to *generate deductions* which recapitulate the reasoning processes of a relevant expert.

exponentiation the mathematical operation of raising one number to the power of another, of which squaring and cubing are examples.

formal system a collection of *effectively* distinguishable *states* and a collection of *rules* which operate *effectively* on states to *generate* other states. Board games such as chess are paradigm *formal systems*.

function a mathematical correlation between some fixed number of inputs and some unique output.

functionalism the philosophical theory of mind according to which the defining features of mental states are their functions in mediating relations between inputs, outputs and other mental states.

generation the process of applying a *rule* to a *state* of a *formal system* to yield another state.

generation tree   a structure showing sequences of applications of *rules* to *states* of a *formal system*.

generative grammar   a *formal system* for ruling on the grammaticality of strings of a language.

Gödel coding   a method involving the *exponentiation* of *prime numbers* which facilitates reference to elements of *formal systems* from within the system.

heuristic (function)   informally, a method for guiding to a solution. Formally, *heuristic functions* assign a value to *states* of a *formal system* representing the closeness of that state to some goal state.

heuristic search   a search method which is informed by a *heuristic function*.

homonym   a *homophone* where multiple semantic interpretations also share an orthographic representation – 'bank' and 'bank'.

homophone   a sequence of phonemes which admits of more than one *semantic* interpretation – 'bred' and 'bread'.

identity theory (of mind)   another name for *Australian/reductive/central state materialism*.

iff   if and only if.

induction   the form of inference characteristic of the empirical sciences, such that the truth of the premises lends support to the conclusion but does not guarantee it.

intentionality   a technical philosophical term for the property of mental states such that they represent something or are about something.

inter alia   among other things.

ipso facto   by that very fact.

isomorphism   an *isomorphism* from one *formal system* to another is a structure-preserving uniform substitution of elements of the system such that all formal properties are thereby preserved.

knowledge argument   an argument, drawn from a thought experiment, which seeks to show that a complete physical explanation of mentality does not constitute a complete explanation.

Loebner Prize   an instituted and annually conducted implementation of the *Turing test*.

logic   the research programme which investigates *logics*.

logics   *formal systems* which encode consequence relations – what 'follows from' what as a matter of logical form.

materialism, reductive/Australian/central state   various names for the philosophical theory of mind according to which types of mental states are identical to types of neural states.

minimax   a procedure for determining winning strategies in the context of a *formal system* representing a two-player game.

modus ponens   the logical principle according to which we can *deduce*, from the truth of the *antecedent* of a *conditional*, the truth of its *consequent*.

modus tollens   the logical principle according to which we can *deduce*, from the falsity of the *consequent* of a *conditional*, the falsity of its *antecedent*.

monophthong   a *sonorant phoneme* whose production involves holding the tongue at one of the cardinal vowel positions.

multiple realisability   with respect to mental states, the fact that they can have more than one physical implementation, either across subjects or within a subject across time.

neural plasticity   the property of human brains such that certain parts of the brain can take up the customary function(s) of damaged parts of the brain – particularly prevalent in younger brains.

neuron   an individual cell of the nervous system.

normative   broadly speaking, a principle, or set of principles, which purports to tell us what we ought to do.

obstruent   a *phoneme* whose production requires that the passage of air through the articulatory apparatus be partially or completed obstructed.

occasionalism   a kind of mind–body *dualism* according to which minds and bodies do not interact, but rather God steps in from time to time to make things seem as if they do.

Ockham's razor   a methodological constraint on theory construction which maintains that one shouldn't postulate any more entities than are strictly explanatorily necessary.

ontology   the core of metaphysics which deals with what, fundamentally, there is – the ontology of *dualism* is one according to which there are two distinct substances.

operant conditioning   the kind of conditioning described by Skinner, which differs from *Pavlovian* conditioning in that the stimulus following the behaviour also has a conditioning effect.

parallelism   a kind of mind–body *dualism* according to which minds and bodies do not interact, but God set things up initially – in preordained harmony – in such a way that it appears as if they do.

Pavlovian conditioning   the process by which some particular stimulus comes to reliably give rise to some particular behaviour.

per se   as such.

petitio principii   the fallacy of begging the question, wherein one assumes the concession of the very proposition in question.

phonemes   idealised categories – to which *phones* assimilate – which represent the stock of atomic meaningful speech sounds in a language.

phones   phonetically realised speech sounds.

phonetic realisation rules   the rules according to which *allophonic* variants of *phonemes* are produced in context-dependent complementary distributions.

phrase structure grammar   a *deterministic formal system* such that each *rule* is context free and contains exactly one symbol on the input side.

phrase structure tree   the *generation tree* of a phrase structure grammar.

positivism   the false doctrine, influential in the late nineteenth and early twentieth centuries, according to which 'real' or 'positive' science should treat exclusively of observable entities.

predicate   in linguistics, a verb with all its complements. For the purposes of this volume a *predicate* can be thought of as expressing a property or relation.

prima facie   on the face of it – on first inspection.

prime number   a number whose only factors are itself and one.

program   the rule which governs the operations of a register machine.

qualia   a technical philosophical term for the subjective, qualitative, privileged, first person aspects of experience.

recursive definition   a means of finitely specifying an infinite class, consisting of a base clause and a recursive clause. The binary numbers can be defined recursively as follows: one and zero are both binary numbers (base clause) and adding a one or a zero to any binary number results in a binary number (recursive clause).

reflex arc   a *Pavlovian* term which is intended to account for the relation between stimulus and behaviour.

register machine   a kind of *formal system* which can be used to define *computation*.

rules   one of the two collections which constitute a *formal system*. For any given *rule* and any given *state* it must be *effective* whether or not the rule applies and, if it does, it must *effectively* deliver a finite number of output states.

semantics   broadly speaking, meaning.

sonorant   a *phoneme* whose production requires that the passage of air through the articulatory apparatus not be obstructed.

states   one of the two collections which constitute a *formal system*. *States* can be defined over any collection of entities, provided there is an *effective* procedure for distinguishing between any two states.

straw man fallacy mischaracterising an opposing position as being weaker than it actually is, then arguing against the weaker position.

substrate independence the ability to be realised in any substrate – it is a property of functionalist theories of mind that mental states are held to be *substrate independent*.

symmetrical a property of relations such that for all A and B, if A stands in the relation to B then B stands in that relation to A. An example is 'being a sibling of'.

synapse a connection between *neurons*.

syntax the rule governed combination of symbols.

terminal state a *generated state* of some *formal system* such that no *rules* of the system apply to it.

token physicalism the philosophical theory of mind according to which, whenever a subject is in a mental state, they are in some neural state but no identification is made between types of mental and neural states.

transduction the process by which an electrical signal is converted to a chemical signal and *vice versa*.

transfer function one of the two functions implemented by each processing node of an *artificial neural network*, which determines what a node's level of activation should be given its afferent activation (and possibly its antecedent level of activation).

transitive a property of relations such that for all A, B and C, if A stands in the relation to B and B stands in the relation to C, then A stands in that relation to C. An example is 'being taller than'.

Turing test a test involving human and computer participants and a human interrogator, such that if the computer can deceive the human interrogator into believing it is human, we should be prepared to say of the computer that it has a mind.

universal machine a particular kind of *register machine* which can, by virtue of *Gödel coding*, take any register machine program as input and operate that register machine.

validity a property of logical forms of inference, such that the truth of the premises guarantees the truth of the conclusion – *valid* inferences admit of no *counter-examples*.

wetware refers to the brain in its capacity as information processing hardware.

INDEX